WONDERS IN MOTION

THE STORY OF DYNAMICS

DAVID ACHESON

WONDERS IN MOTION

THE STORY OF DYNAMICS

OXFORD
UNIVERSITY PRESS

OXFORD
UNIVERSITY PRESS

Great Clarendon Street, Oxford, OX2 6DP,
United Kingdom

Oxford University Press is a department of the University of Oxford.
It furthers the University's objective of excellence in research, scholarship,
and education by publishing worldwide. Oxford is a registered trade mark of
Oxford University Press in the UK and in certain other countries.

Published in the United States of America by Oxford University Press
546 Fifth Avenue, New York, NY 10036, United States of America

British Library Cataloguing in Publication Data
Data available

Library of Congress Control Number: 2026938403

ISBN 9780198996859

DOI: 10.1093/oso/9780198996859.001.0001

Printed and bound by
CPI Group (UK) Ltd, Croydon, CR0 4YY

FSC www.fsc.org MIX Paper | Supporting responsible forestry FSC® C013604

The manufacturer's authorized representative in the EU for product safety is
Oxford University Press España S.A. of Parque Empresarial San Fernando de Henares,
Avenida de Castilla, 2 – 28830 Madrid (www.oup.es/en or product.safety@oup.com).
OUP España S.A. also acts as importer into Spain of products made by the manufacturer.

Contents

1. Introduction 1

2. The Beginning of Dynamics 5

3. Force, Mass, and Acceleration 12

4. Circular Motion 18

5. Energy 23

6. Collisions 28

7. 'A Scab of Symbols' 34

8. Reaching for the Heavens 40

9. *Principia* 47

10. Newton and Planetary Motion 52

11. Calculus 57

12. Oscillations 63

13. A New Age Begins... 67

14. ‘Contrary to Y[e] Opinion of Y[e] Vulgar’ 73

15. The Principle of Linear Momentum 77

16. The Principle of Angular Momentum 80

17. Spinning Tops 87

18. . . . and Falling Cats 93

19. The Not-so-Simple Pendulum 99

20. The Wave Equation 104

21. ‘The Lawes . . . of Springinesse’ 111

22. The Best of All Possible Worlds? 116

23. Instability 122

24. Fluid Dynamics 127

25. Viscous Flow 133

26. ‘I Got Instantly on Horseback’ 141

27. Shock Waves 148

28. Vortex Motion 154

29. The Curse of the Golden Ratio 160

30. The Coriolis Force 166

31. The Three-Body Problem 171

32. A Paradox Resolved 177

33. The Secrets of Flight 183

34. Falling through the Earth 190

35. Tremors in the Night 198

36. Very Viscous Flow 204

37. Which Way Does the Bathwater Spin? 209

38. Tipping Points, Catastrophe, and Chaos 216

39. And Finally . . . 223

Notes 229
Acknowledgements 257
Figure Acknowledgements 258
Index 260

1

INTRODUCTION

You are on the inside of an enormous rotating drum, pinned against the wall by some 'force' that doesn't actually exist.

The noise is deafening, and, to make matters even worse, the floor is gradually receding beneath your feet. Yet you don't fall down . . .

What on earth is going on?

Fig. 1 Defying gravity?

In truth, dynamics is never easy. It took some extremely clever people a very long time, for instance, to realize that force is not associated directly with motion, but with *changes* in motion, especially through the law which we express today as

$$Force = Mass \times Acceleration.$$

These ideas were pioneered by Sir Isaac Newton, in his *Principia* of 1687, most famously in connection with the motion of the planets.

Yet it was not until some 60 years later that classical mechanics came to be done in much the way it is done today, largely through the work of the great Swiss

(a) (b)

Fig. 2 (a) Isaac Newton (1642–1727). (b) Leonhard Euler (1707–1783).

mathematician Leonhard Euler, who also broadened the whole subject to include rigid body dynamics, elasticity, and fluid mechanics (Fig. 3).

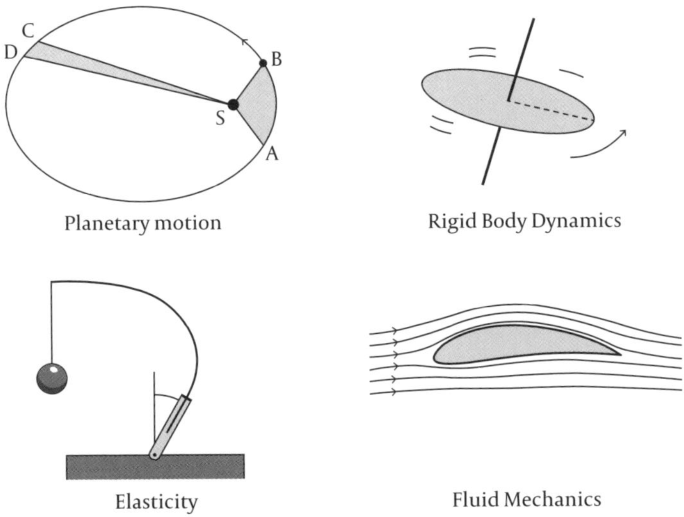

Fig. 3 Some elements of classical mechanics.

It was largely through Euler's work, too, that a further principle of mechanics gradually emerged—one concerned more directly with *rotation*.

* * * * * * *

This book tells the story of classical mechanics, right up to the present day.

The major focus throughout will be on the actual physical phenomena involved, with particular emphasis on those that contain an element of *surprise* and most challenge our intuition.

Nonetheless, the development will be roughly chronological. And while it is possible to find a handful of valuable ideas much earlier, our story begins, really, with Galileo . . .

2

THE BEGINNING OF DYNAMICS

It is an iconic scene: Galileo, in around 1590, dropping two different weights from the Leaning Tower of Pisa, to show that they hit the ground at the same time.

But did it really happen?

Fig. 4 Myth or reality?

The experiment had, in fact, been performed a few years earlier, by Simon Stevin in Holland, who recorded taking two balls of lead, one ten times heavier than the other, and dropping them through a height of 30 feet at the Nieuwe Kerk in Delft. But whether Galileo himself performed the experiment is not clear, for there is no record of it in his own working papers.

We do know, however, that he conducted other experiments on the way in which objects descend under gravity, the most famous being one that involved motion down an inclined plane . . .

The inclined plane experiment

Galileo's plane made only a few degrees with the horizontal, greatly slowing the whole descent so that it was more susceptible to precise measurement.

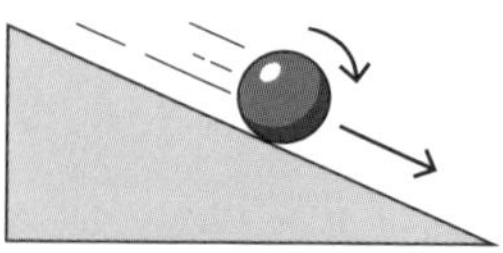

Fig. 5 Galileo's inclined plane experiment.

And in his *Two New Sciences* of 1638 he gave a detailed account of the whole procedure:

> A piece of wooden moulding or scantling, about 12 cubits long, half a cubit wide, and three finger-breadths thick, was taken; on its edge was cut a channel a little more than one finger in breadth . . . we rolled along it a hard,

> smooth and very round bronze ball. Having placed this board in a sloping position, by lifting one end some one or two cubits above the other, we rolled the ball . . . along the channel, noting . . . the time required to make the descent.

Fig. 6 Galileo Galilei (1564–1642).

Timing the descent was, in fact, no easy matter, for Galileo had no clock accurate enough for the purpose, so had to devise a method of his own:

> For the measurement of time, we employed a large vessel of water placed in an elevated position; to the bottom of this vessel was soldered a pipe of small diameter giving a thin jet of water, which we collected in a small glass during the time of each descent . . . the water thus collected was weighed . . . on a very accurate balance; the differences and

> ratios of these weights gave us the differences and ratios of the times . . .

And Galileo summarized the major conclusion from all these experiments as follows:

> . . . we always found that the spaces traversed were to each other as the squares of the times . . .

In other words, if we let t denote the time since the start of the motion, Galileo found that the distance travelled was proportional to t^2.

So, after one unit of time the ball travels a certain distance, but after two units of time the total distance travelled is not twice as far but four times as far, because $2^2 = 4$.

The ball therefore *accelerates*, and this immediately raises some deep questions about what its speed down the plane *is*, exactly, at any given moment . . .

Two simple types of motion

In the earliest days of dynamics, there were really only two types of motion in a straight line that could be handled at all successfully.

Motion at constant speed

In this simplest of cases, the distance travelled is proportional to the time taken, and the speed itself is then given by the well-known formula

$$speed = \frac{distance}{time}.$$

Moreover, if we plot the distance travelled against time, we get a straight line, and the speed itself is then simply the slope of the line (Fig. 7).

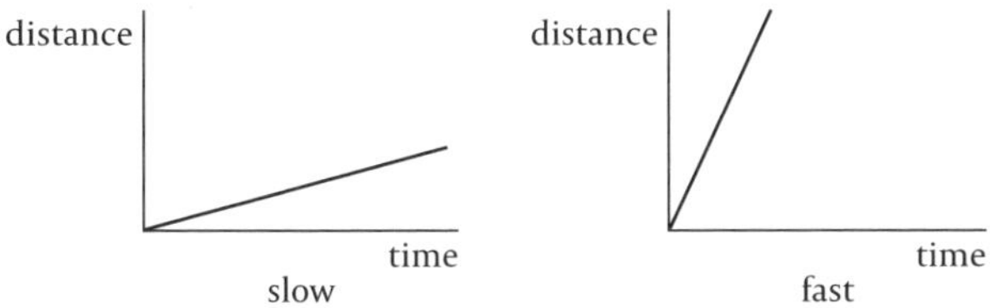

Fig. 7 Motion at constant speed.

It is important to emphasize, however, that these simple and well-known ideas apply directly only when the speed is *constant.*

And in Galileo's experiment the speed at which the ball travels down the inclined plane clearly increases with time . . .

Motion with constant acceleration

If we let v denote the speed down the inclined plane at time t, Galileo found, in fact, that

$$v = at,$$

where a is a constant which we now call the *acceleration.*

Moreover, he concluded that the distance travelled down the plane, from an initial state of rest, was

$$s = \frac{1}{2}at^2.$$

Both these aspects of motion with constant acceleration had, in fact, been discovered in the fourteenth century by scholars of Merton College, Oxford, albeit in an abstract way that appears to be detached from any experiment. And a graphical interpretation of their main result had subsequently been given by Nicole Oresme, in Paris (Fig. 8).

A loose paraphrase of Oresme's argument can be seen in Fig. 9, where we consider first the case of constant speed. For

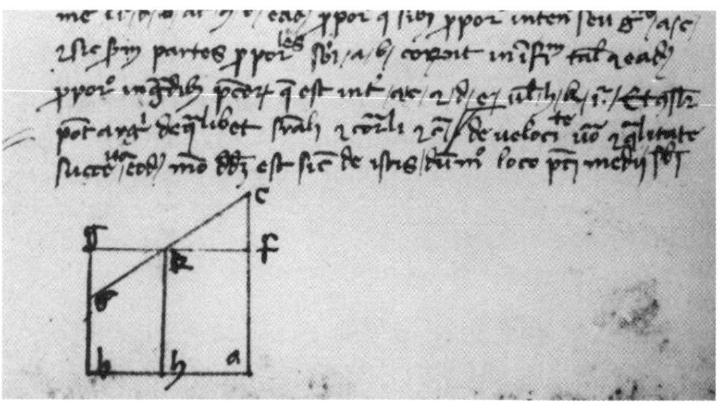

Fig. 8 From a fifteenth-century manuscript of Oresme's *De configurationibus qualitatum*, MS Florence, Biblioteca Nazionale, Conv., Suppr., J. Ix, 26, f. 32^v.

if we plot the graph of speed v against time t, it is immediately evident that the distance travelled, vt, corresponds to the area under the graph (Fig. 9a).

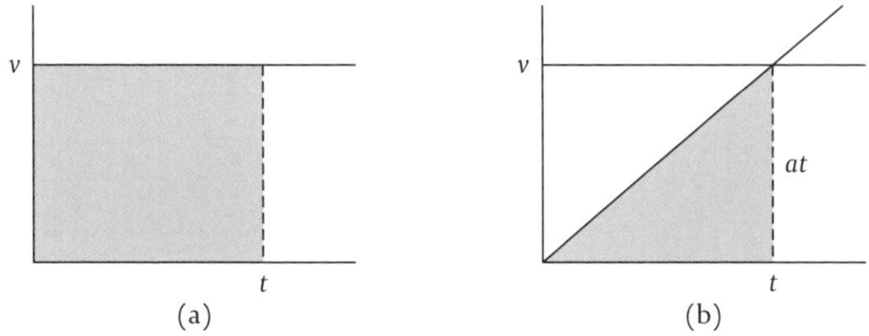

Fig. 9 Motion (a) at constant speed and (b) from rest, with constant acceleration a.

So, if we take the same approach to the uniformly accelerated motion from rest in Fig. 9b, the distance travelled will be the triangular area shown, which is indeed $\frac{1}{2}at^2$.

Most importantly of all, perhaps, Galileo came to the conclusion that an object falling freely under gravity does so in exactly the same manner.

And he was right: in the absence of air resistance, all objects near the Earth's surface do indeed fall under gravity with the same constant acceleration:

$$g = 9.81 \text{ m/s}^2.$$

But *why*?

3

FORCE, MASS, AND ACCELERATION

Today, we accord each object a certain *mass* m, and regard its fall as due to a downward gravitational force mg, in keeping with the idea that

$$force = mass \times acceleration.$$

Fig. 10 The force due to gravity.

To proceed much further, however, we need to take a closer look at force in general, which is a *vector* quantity, with both magnitude and direction.

Force as a vector

It makes sense, I think, to first consider dynamics in just two dimensions.

Any force can then be split into two *components* in the directions shown in Fig. 11a, and its dynamical effect is completely equivalent to the combined effects of those two components.

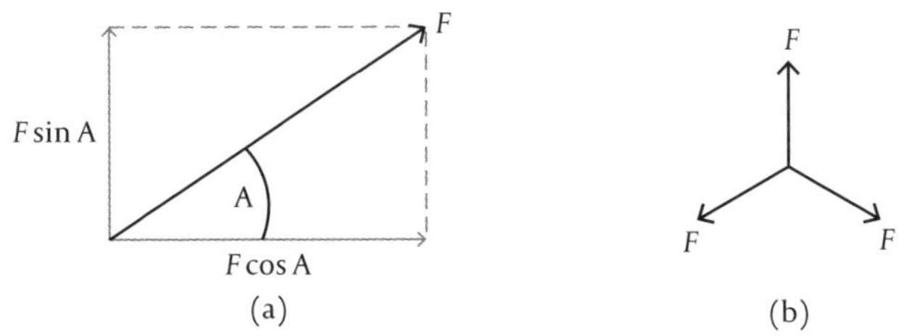

Fig. 11 Force as a vector.

A simple example is shown in Fig. 11b, where three equal forces F act at a point and at 120° angles to one another. By symmetry, they must clearly balance, and if we consider for a moment just their vertical components, then the two lower ones are both $F \cos 60^\circ$ which is, indeed, $F/2$, just as required.

As it happens, Simon Stevin invented a much more ingenious 'thought experiment' to demonstrate the whole idea in 1586, and a crude paraphrase is shown in Fig. 12b, where a uniform chain hangs over an object.

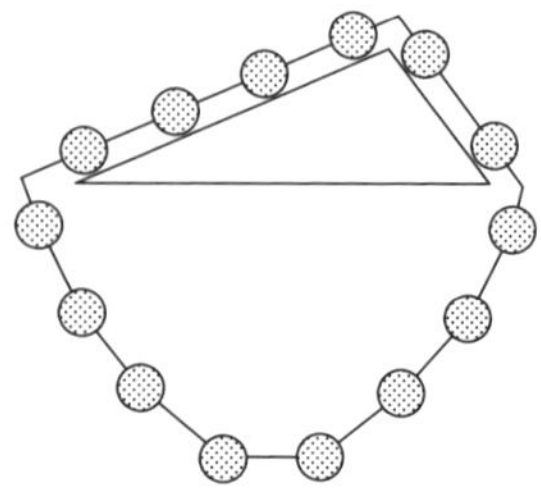
(a)

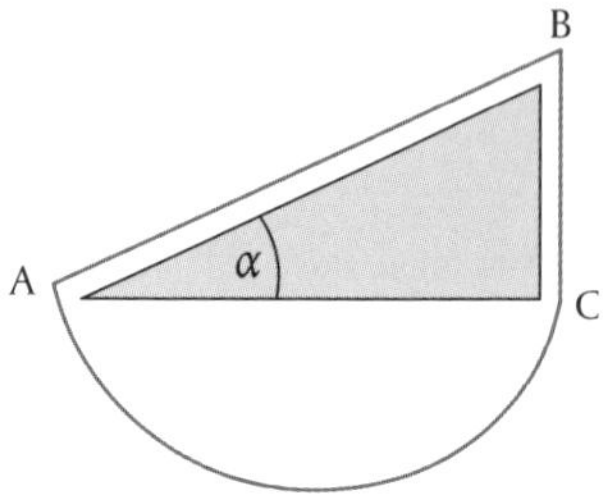

(b)

Fig. 12 (a) Detail from Stevin's *Beghinselen der Weeghconst* (1586). (b) A paraphrase of Stevin's argument.

Stevin argued that the chain must be in equilibrium, or perpetual motion would result, which he deemed absurd. He also argued that the portion of the chain below the line AC would hang symmetrically, and could therefore be removed without disturbing the equilibrium.

Finally, the portion BC is lighter than the portion AB by a factor of $BC/AB = \sin\alpha$. So, in order for the two to be in balance, if the weight of the portion AB is mg, then the component of that weight acting *down the plane* must be only $mg \sin\alpha$.

The inclined plane

In view of what has just been said, we might expect the acceleration down the plane in Galileo's famous experiment to be $g \sin \alpha$. And it would be if the plane were frictionless, in which case the ball would simply slide down.

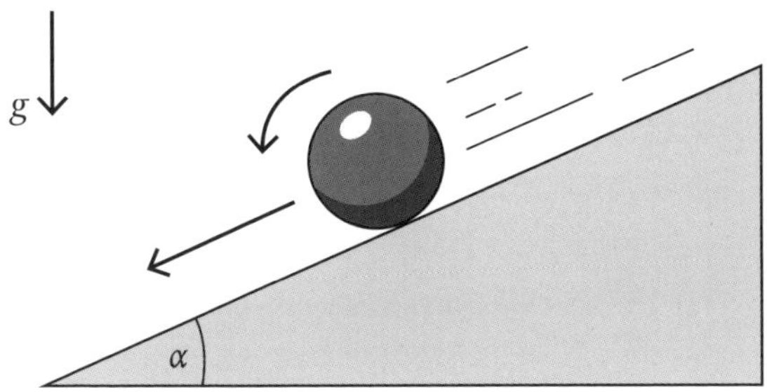

Fig. 13 Galileo's inclined plane revisited.

Instead, however, the ball *rolls* down, and the acceleration of its centre down the plane is then reduced to $\frac{5}{7} g \sin \alpha$, for subtle reasons that we will see later in the book (Notes, p. 230).

Nonetheless, this acceleration down the plane is still *constant*, and still proportional to $\sin \alpha$.

And this allows us to explain a rather elegant theorem . . .

Galileo's chord theorem

As one might expect, Galileo experimented with planes inclined at different angles to the horizontal, and he discovered a remarkable result which unified them all.

In his *Two New Sciences* of 1638, Galileo states this theorem as follows:

> If from the highest or lowest point in a vertical circle there be drawn any inclined planes meeting the circumference the times of descent along these chords are each equal to the other.

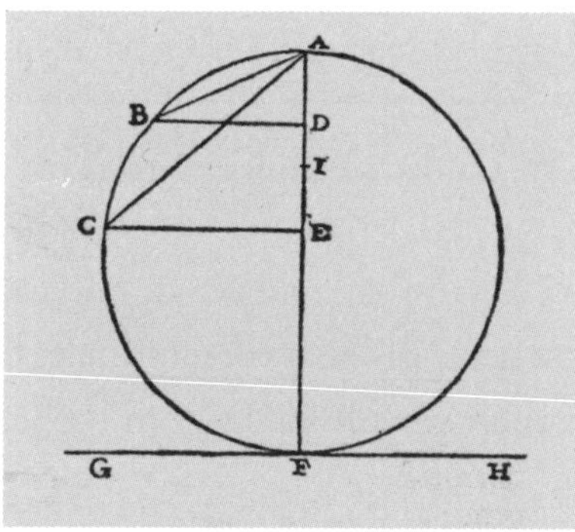

Fig. 14 Galileo's sketch for his 'chord theorem'.

A ball rolling from A to B in Fig. 14 therefore takes exactly the same time as one rolling from A to C. In the latter case, the slope is steeper, but the ball also has further to go, and the two effects exactly cancel out (Notes, p. 230).

* * * * * * *

Galileo also conducted some altogether different experiments on how gravity affects the free motion of an object when it is projected horizontally.

And he found, in particular, that the result was simply a combination of the original horizontal motion at constant speed and free fall under gravity, giving rise to motion in a parabola (Fig. 15).

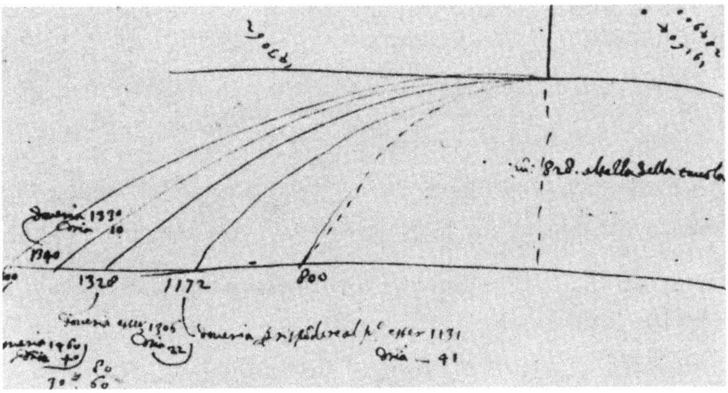

Fig. 15 Galileo's sketch of his experiments on parabolic projectile motion, from folio 116^{v} of his working papers, now kept in the Biblioteca Nazionale Centrale, Florence.

Yet, as it happens, it was a rather different type of motion that would eventually propel the subject even further . . .

4

CIRCULAR MOTION

Any reader with an occasionally misspent youth will know that if you whirl an object in a sling you have to exert a force *inward* to stop it flying off at a tangent.

And this is wholly consistent with the idea of *force* = *mass* × *acceleration*, because the object in question is indeed accelerating all the time towards the centre of the circle (Fig. 16).

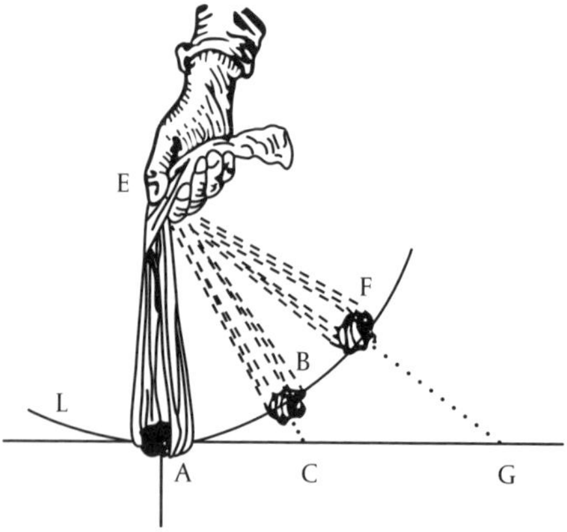

Fig. 16 From René Descartes' *Principles of Philosophy* (1644).

In case this seems odd at all, it may be because in dynamics the word 'acceleration' has a very different meaning to the one it tends to have in everyday life . . .

Velocity and acceleration

I should first clarify the distinction in mathematics and science between speed and velocity. For speed is simply a number, with certain units, while velocity is a *vector* quantity having both magnitude *and direction* (Fig. 17).

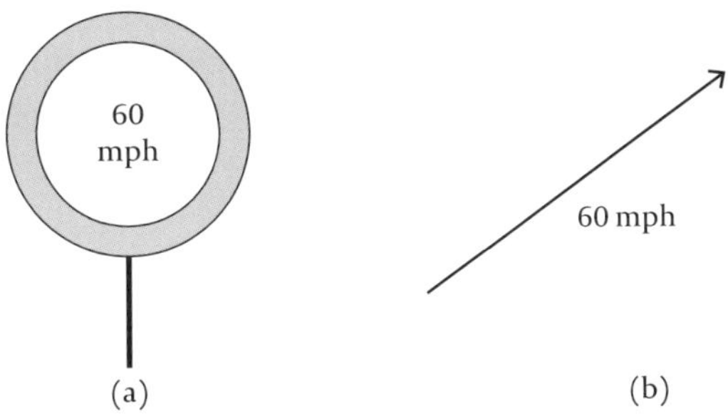

Fig. 17 The distinction between (a) speed and (b) velocity.

Now, in everyday life, acceleration tends to mean rate of change of speed, without regard to direction of motion. In dynamics, however,

acceleration = rate of change of velocity.

And in circular motion, even when the magnitude of the velocity is constant, its *direction* is changing all the time.

Acceleration in circular motion

Suppose, then, that an object is moving round a circle of radius r at constant speed v. To determine its acceleration we loosely paraphrase an argument due to Newton in 1669.

In Fig. 18, let the object be at point A at $t = 0$, and therefore moving to the left with speed v. In the absence of a force, it would 'fly off at the tangent', and a very short time t later it would be at the point B.

Instead, however, it moves round the circle by a very small arc of length vt, and this brings it very close to the point where the line OB meets the circle. In doing this it therefore 'falls' a very small distance h towards the centre of the circle, O.

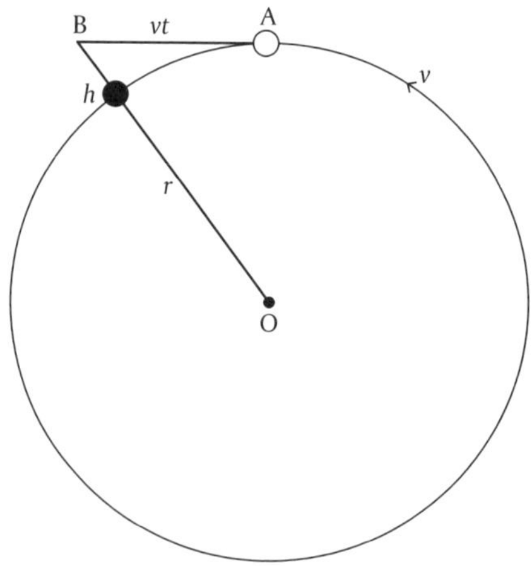

Fig. 18 Finding the acceleration towards the centre.

Newton then cites a theorem from elementary geometry to conclude that $h\,(2r+h) = (vt)^2$ (see Notes, p. 231). And because h is very small compared to $2r$, this implies that, to a very good approximation:

$$h = \frac{1}{2}\frac{v^2}{r}t^2.$$

And this is just the distance formula $\frac{1}{2}at^2$ from Chapter 2 describing motion with constant acceleration a, but with $a = v^2/r$.

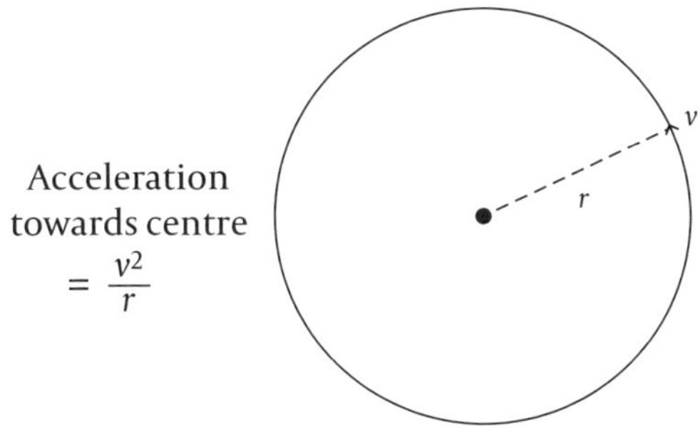

Fig. 19 The acceleration in circular motion.

The centripetal force

At this point, we can actually make sense of what is happening with the fairground rotor of Fig. 1.

First, however, I should mention the fairly common idea that the people and objects there are somehow pinned against the wall by a radially *outward* 'centrifugal force'.

I will explain much later, in Chapter 30, how, in certain circumstances, it can be entirely rational for *someone on the wall* to think like this. For a fixed observer, however, watching the rotor and the people go round, it is sheer nonsense.

From a fixed observer's point of view, each person of mass m on the wall has instead, as we have just seen, an acceleration inwards all the time due to their rotation, and this is only made possible by a radially *inward*

$$\textit{centripetal force} = \frac{mv^2}{r}$$

exerted on them by the wall.

And it is the frictional force associated with that which opposes gravity and prevents them from falling down.

5

ENERGY

One of the most exotic examples of circular motion has to be the 'Grand Centrifugal Railway' of Fig. 20, which appeared in Great Windmill Street, London, in 1842.

Fig. 20 A short-lived entertainment from 1842.

According to the advertisement:

> a carriage of nearly 200 lbs., carrying a Lady or Gentleman, descends an inclined plane, passing round a Vertical Circle of 40 feet, turning the head of the individual downwards and feet in the air; the carriage then descends the Circle, rises a second Inclined Plane, and *lands the individual safe.*

But this version of 'loop-the-loop', with its circular track and promise of safety, turned out to be rather short-lived. And it is not difficult to see why, once we have introduced an important new idea: *energy*.

Kinetic and potential energy

Let us begin by considering again the free fall of an object under gravity (Fig. 21), but from a novel point of view.

For any point mass m moving with speed v possesses, on account of its motion,

$$\textit{kinetic energy} = \frac{1}{2}mv^2.$$

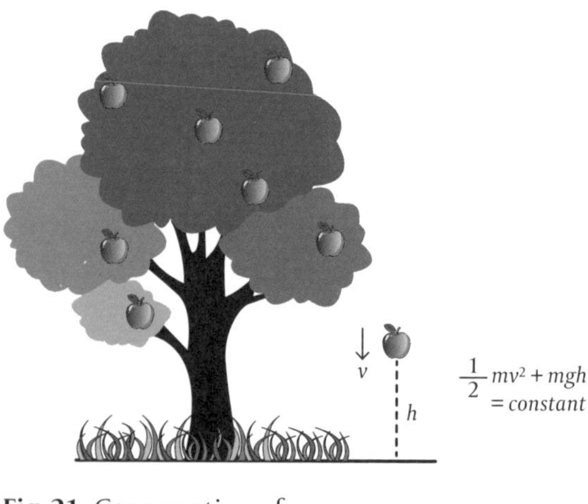

Fig. 21 Conservation of energy.

In a uniform gravitational field it also possesses 'potential energy' on account of its height h:

$$\textit{gravitational potential energy} = mgh.$$

And the result in Fig. 21 expresses the key idea of *conservation of energy.*

It follows, then, that in falling a distance s from a state of rest an object acquires a speed v such that

$$v^2 = 2gs,$$

and this can alternatively be obtained by taking the formulae $s = \frac{1}{2}gt^2$ and $v = gt$, which we noted in Chapter 2, and eliminating time t.

One great merit of working with kinetic energy is that it is fully determined by the *speed* of the object, so that, in this particular respect, the direction of the motion doesn't matter.

And, with much hindsight, this helps us explain another major finding by Galileo.

Another experiment by Galileo

In the course of his experiments with inclined planes, Galileo discovered that the final speed v is the same for all planes with the same initial height (Fig. 22).

And, at first sight, we can understand this readily enough by equating the kinetic energy gained, $\frac{1}{2}mv^2$, with the

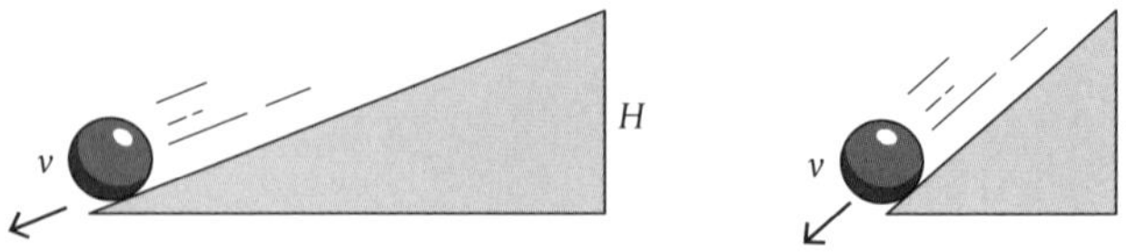

Fig. 22 Another experiment by Galileo.

potential energy lost, mgH, so that $v^2 = 2gH$, irrespective of the plane's angle of inclination.

This is, however, something of an oversimplification, for the argument above is strictly valid for a point mass sliding down a frictionless inclined plane; and the fact that Galileo's spheres were actually rolling introduces, once again, further subtleties.

Loop the loop?

Finally, let us apply all these ideas to the Grand Centrifugal Railway of 1842, by considering first the speed v that is needed at the bottom of the circle to even just *reach* the top.

If the radius of the circle is r, the increase in potential energy between bottom and top will be $2mgr$. The kinetic energy at the bottom, $\frac{1}{2}mv^2$, will therefore need to be at least $2mgr$, so even just reaching the top of the circle requires a speed at the bottom such that $v^2 = 4gr$.

And the trouble with this is that the centripetal acceleration $\frac{v^2}{r}$ at the bottom will then have to be, at the very least, $4g$.

Worse still, it turns out that to stop the occupants simply falling out at the top an acceleration at the bottom of at least 5*g* is needed (Notes, p. 231).

This would have constituted a serious risk of injury for many people, and is why modern loop-the-loop rollercoasters are *not* circular, but carefully designed instead to avoid such extreme accelerations.

6

COLLISIONS

A further idea of major importance in dynamics is

$$\textit{momentum} = \textit{mass} \times \textit{velocity}.$$

And this emerged, very largely, from experiments on collisions.

Fig. 23 A collision problem, from Marcus Marci's *De Proportione Motus* (1639).

In 1666, for instance, the Royal Society of London conducted an experiment in which a wooden ball collided with an identical but stationary one. As a result, the moving one came immediately to rest, while the stationary ball then shot off with the velocity originally possessed by the moving one.

And two years later, in 1668, the society issued a challenge to find the general laws governing such collisions.

One of those to take up the challenge was the Dutch mathematician Christiaan Huygens, and in modern terminology his rules of collision amount to

(i) the total momentum is conserved, and
(ii) the total energy is conserved.

Suppose, for instance, that a mass m, moving with speed u, collides with a stationary mass M, as in Fig. 24. Then the rules given by Huygens amount to

$$mu = mv + Mw$$

and

$$\frac{1}{2}mu^2 = \frac{1}{2}mv^2 + \frac{1}{2}Mw^2.$$

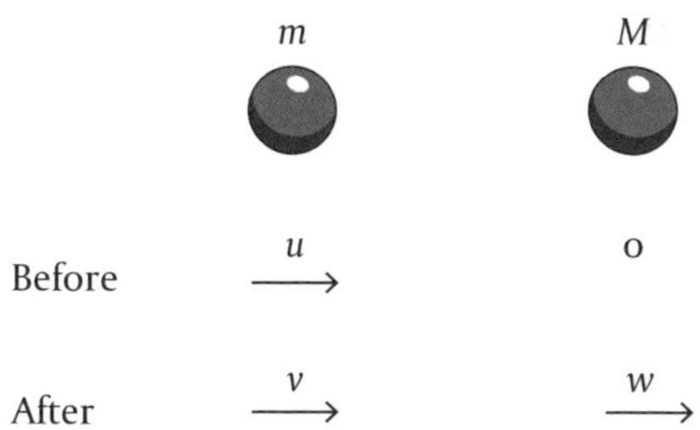

Fig. 24 The velocities before and after a collision.

These two equations suffice to determine the velocities v and w after the collision, which turn out to be

$$v = \left(\frac{m - M}{m + M}\right) u, \qquad w = \left(\frac{2m}{m + M}\right) u.$$

In particular, if the two masses are equal, then $v = 0$ and $w = u$, as in the Royal Society's experiment of 1666. Yet two other special cases leap immediately to mind.

In the first, m is much smaller than M. The incident mass then bounces *back* (because of the minus sign) at almost the original speed—much as if it has hit a brick wall—and the large mass M moves only slowly.

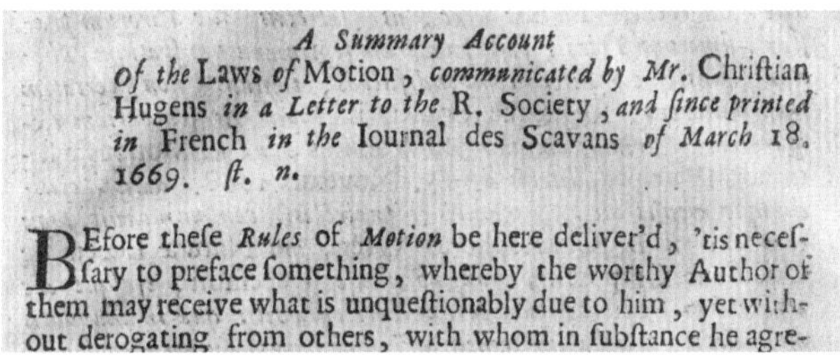

A Summary Account
of the Laws *of* Motion, *communicated by Mr.* Chriſtian Hugens *in a Letter to the* R. Society, *and ſince printed in* French *in the* Iournal des Scavans *of March* 18. 1669. *ſt. n.*

BEfore theſe *Rules* of *Motion* be here deliver'd, 'tis neceſſary to preface ſomething, whereby the worthy Author of them may receive what is unqueſtionably due to him, yet without derogating from others, with whom in ſubſtance he agre-

Fig. 25 The beginning to Huygens' paper in the Royal Society's *Philosophical Transactions*, 12 April 1669.

At the other extreme, when m is much larger than M, the velocity of the incident mass is hardly affected, while the mass M moves off at a speed almost *twice* the original incident speed u.

And at this point I should admit that this last result once revealed a serious breakdown in my own physical intuition.

For, at the time, I had casually imagined that a large enough incident mass could make the smaller mass shoot off more or less as fast as we might like.

And I came to my senses only by looking at the whole problem a completely different way . . .

Galileo on relativity

In one rather colourful passage from Galileo's *Dialogue Concerning the Two Chief World Systems* (1632), he invites us to shut ourselves up in a cabin, deep within some ship, together with some butterflies in the air, some fish in a bowl, and a bottle from which water is slowly dripping. He invites us to observe carefully the various motions of the

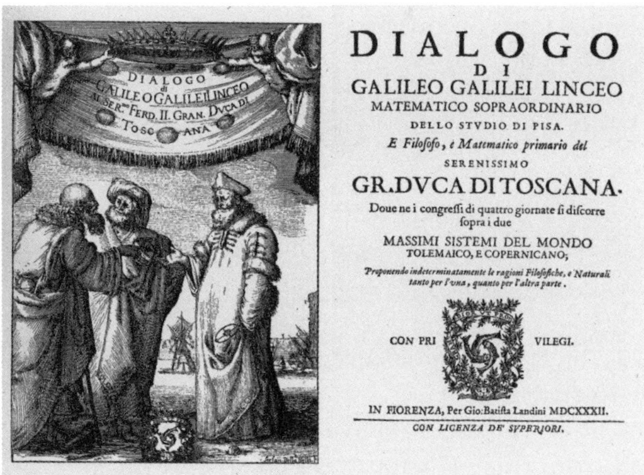

Fig. 26 Galileo's *Dialogo* of 1632.

butterflies, fish, and water, and claims that we would be completely unable to tell from these observations whether the ship was

(a) stationary or
(b) moving at constant velocity.

In modern terms, Galileo asserted that the laws of dynamics are the same with respect to all reference frames which move at *constant* velocity relative to one another.

A problem revisited

Given Galileo's insight, then, we are entitled to view the collision problem in Fig. 24 from a reference frame moving to the right with constant speed u, and yet still apply the same dynamical ideas of conservation of energy and momentum.

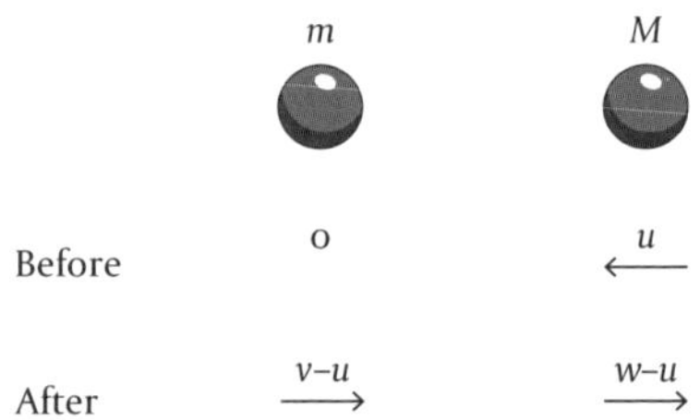

Fig. 27 The collision problem again, but viewed from a different reference frame.

And from this viewpoint, the mass m is initially at rest, while the mass M is incident from the right, with speed u, and the velocities after the collision are as given in Fig. 27.

So, if w were to be greater than $2u$, the mass M would rebound from the collision *at greater speed than it went in.* And this would be absurd, for, having imparted some kinetic energy to the stationary mass m, it would violate conservation of energy.

7

'A SCAB OF SYMBOLS'

We have now met several of the most fundamental concepts of dynamics, though only by occasionally playing rather fast and loose with the history of the subject, especially regarding ideas of energy, which developed fairly late.

It is easy to forget, too, that the whole idea of *algebra*—at least as we know it today—was still in its infancy in the

Fig. 28 Wallis's *Treatise of Algebra* (1685).

seventeenth century. And it was not until about 1685, with books like John Wallis's treatise on the subject (Fig. 28), that things really began to change.

Even then, however, algebra did not have an easy ride, and Wallis's earlier writings on the subject had been famously dismissed by the philosopher Thomas Hobbes as

> a scab of symbols.

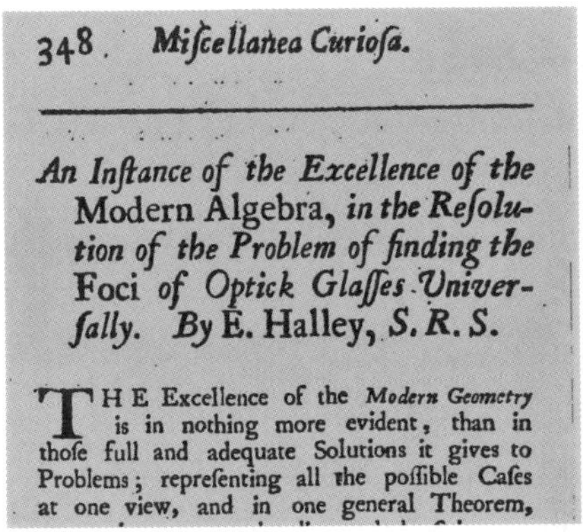

348. *Miſcellanea Curioſa.*

An Inſtance of the Excellence of the Modern Algebra, *in the Reſolution of the Problem of finding the* Foci *of Optick Glaſſes Univerſally.* *By* E. Halley, *S. R. S.*

THE Excellence of the *Modern Geometry* is in nothing more evident, than in thoſe full and adequate Solutions it gives to Problems; repreſenting all the poſſible Caſes at one view, and in one general Theorem,

Fig. 29 Halley's paper on optics of 1693.

The whole idea of an algebraic *formula*, to describe some aspect of physics, was even slower to evolve. As late as 1693, and well after Newton's *Principia*, we find a paper on optics by Edmund Halley extolling the virtues of 'modern' algebra on the grounds of its generality (Fig. 29), or—as Halley puts it—its ability to represent

> all possible Cases at one view.

And it isn't just the way that algebraic formulae capture general ideas that gives them such power in the study of dynamics. There is another aspect, too, which came into its own only much later . . .

The method of dimensions

In any problem of pure mechanics, the physical dimension of each quantity must be some combination of mass (M), length (L), and time (T).

Velocity, for instance, has dimension of L/T, being measurable in metres per second.

Acceleration, being rate of change of velocity, has dimension of L/T^2. (The acceleration g due to gravity, for instance, is 9.81 m/s^2.)

Force has dimension ML/T^2, essentially because of the fundamental law *force* = *mass* × *acceleration*.

And these ideas can often provide simple but effective checks on progress in mechanics. In Chapter 4, for instance, we found that the acceleration towards the centre in circular motion was v^2/r. As the speed v has dimension L/T and the radius r has dimension L, this does indeed have dimension L/T^2, as required. And, if, for instance, I had made a slip at the time, and claimed instead that the acceleration in circular motion was v/r, you would have been able to tell instantly that something was wrong.

Dimensions	
Mass	M
Length	L
Time	T
Velocity	L/T
Acceleration	L/T^2
Momentum	ML/T
Force	ML/T^2
Energy	ML^2/T^2

Fig. 30 Dimensions in mechanics.

Even more remarkably, however, dimensions can, on occasion, offer much more than just a check . . .

The simple pendulum

Consider, for instance, one of Galileo's most famous discoveries: that each to-and-fro oscillation of a simple pendulum takes place in a time which is proportional to the *square root* of its length, so that longer pendulums oscillate more slowly (Fig. 31).

However odd it might seem, the method of dimensions allows *us* to deduce this in the following way, without any experiment at all.

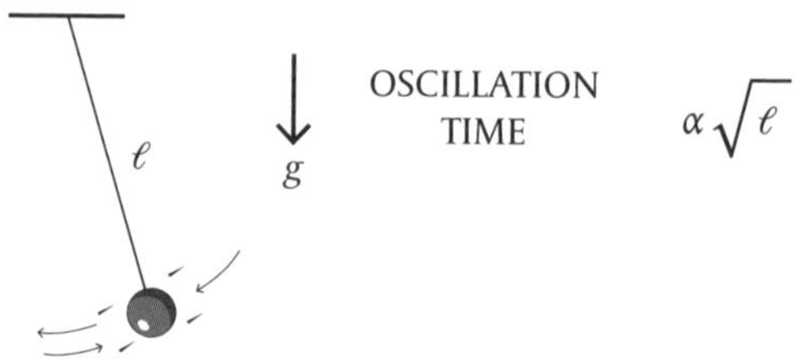

Fig. 31 The simple pendulum.

For what physical quantities can possibly be relevant here? The pendulum length *l*, with dimension L, would certainly seem relevant, and it is only natural to suppose that the mass *m* will play a role, with dimension M. Finally, the whole phenomenon is clearly a gravitational one, so we must expect *g* to appear, with dimension L/T^2.

In the absence of other physical effects, such as air resistance, no other quantities seem relevant. So, as we are interested in the oscillation *time*, we need to create from just these three quantities something with dimension T.

And it is at once apparent that we cannot, in fact, involve the mass of the bob *m* at all, and that the only way of creating something with dimension T is to divide *l* by *g*, to get something with dimension T^2, and then take the square root.

In this way, we learn *from dimensional considerations alone* that the oscillation time of a simple pendulum must be proportional to $\sqrt{l/g}$, and hence proportional to the square root of its length.

And in case this appears little short of magical, and we are left wondering 'Who really did all the work?', the answer, in short, is Galileo.

For the whole argument rested crucially on knowing that gravity near the Earth's surface is properly represented by what we now call an *acceleration*, g, rather than some other physical quantity having quite different dimensions.

* * * * * * *

Armed with all these ideas, I would now like to return to the main story, and show how dynamics really began to flourish in the age of Newton, largely because attention turned towards the heavens . . .

8

REACHING FOR THE HEAVENS

Sometimes in mathematics and science, an old idea gets a completely new lease of life, long after it was first conceived. And one of the most notable examples is the *ellipse* (Fig. 32).

It began life, in ancient Greece, with several interesting geometrical properties. For while an ellipse is, in one sense, just a cross-section of a cone, Fig. 32 also shows a quite different way of constructing one.

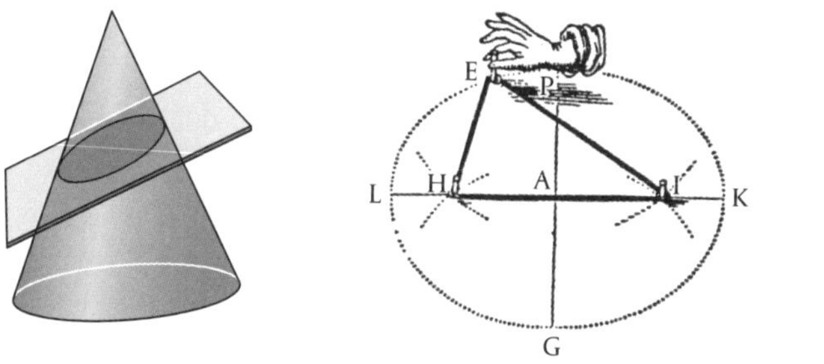

Fig. 32 Properties of an ellipse.

All we have to do is mark out two points H and I, and thread a loop of string round them. If we then move the point E, while keeping the string taut, we trace out an ellipse.

More remarkably still, it turns out that the two lines HE and IE make equal angles with the tangent to the ellipse at E. This means that with an elliptical mirror, a beam of light from H to E will get reflected to I, and that is why the two points H and I are called the *focal points* of the ellipse.

And while all this was known in ancient Greece, the ellipse was destined to acquire a wholly new significance much, much later . . .

* * * * * * *

In 1609, the German astronomer Johannes Kepler published a careful analysis of the orbital data for the six planets that were known at the time, and summarized his conclusions in two famous laws.

First, each planet moves in an elliptical orbit, with the Sun S not at the centre of the ellipse, but at one of its *focal points* (Fig. 33).

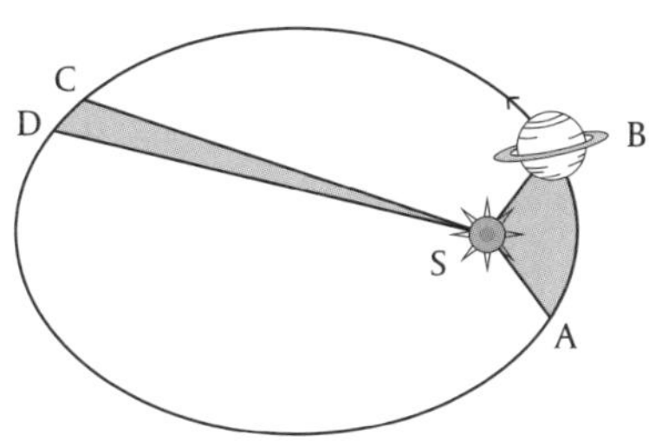

Fig. 33 Planetary motion. The shaded areas here are equal, and the planet takes as long to get from C to D as it does to get from A to B.

Second, a line drawn from the Sun to the planet *sweeps out equal areas in equal times*, so that the planet moves faster when it is closer to the Sun.

And ten years later, in 1619, he followed this with a third law, which concerned the different orbital *times* of the various planets. For if T denotes the time for each orbit, and R the planet's mean radius from the Sun, then

$$T^2 \textit{ is proportional to } R^3.$$

Thus Jupiter, for instance, which is over five times further from the Sun than the Earth, has an orbital period of almost 12 Earth years.

And yet the obvious question arises . . . what *causes* all this?

* * * * * * *

By 1666 the problem was beginning to attract the attention of the Royal Society, and in the records for May 9th that year we find

> It being mentioned by Mr. Hooke that the motion of the celestial bodies might be represented by pendulums, it was order'd, that this should be shewed at the next meeting.

Hooke's idea was the so-called conical pendulum (Fig. 34a), for when this makes a small angle with the vertical, the bob typically moves along a path which is very nearly an ellipse. And Hooke rightly perceived that the horizontal component of force on the bob, due to the supporting string, is always directed towards O (Fig. 34b).

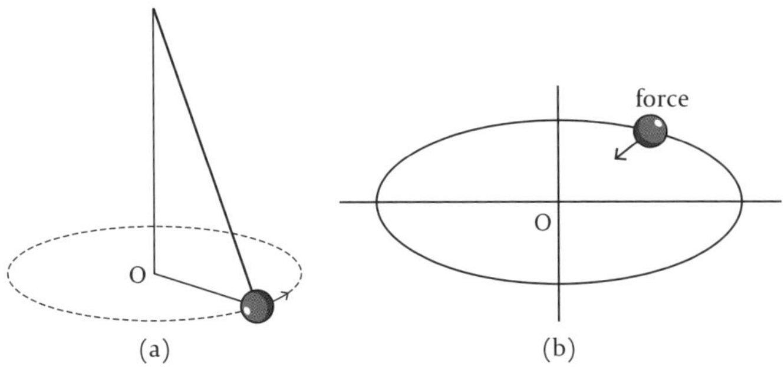

Fig. 34 (a) Motion of a conical pendulum and (b) the horizontal component of force due to the string.

Now, this pendulum analogy with planetary motion was far from perfect, not least because the point O here lies at the centre of the ellipse rather than at one of its focal points.

Nonetheless, Hooke wondered if something similar might be operating in the heavens, with an attractive force towards the Sun continually drawing each planet from what would otherwise have been a straight-line path.

A simple calculation

Seven years later, in 1673, Huygens' *Horologium Oscillatorum* was published, containing (in verbal form) the formula v^2/r for the centripetal acceleration in circular motion.

This at once allowed a most important calculation, because the planetary motions are only slightly elliptical.

So, if we approximate them by *circles* of radius r, we can use Kepler's third law to work out how the speed v, and therefore v^2/r, depends on r.

For the time taken for each complete orbit will then be circumference divided by speed, $2\pi r/v$. Kepler's third law therefore implies that r^2/v^2 is proportional to r^3, so that v^2 must be proportional to $1/r$. This, in turn, implies that v^2/r, the acceleration towards O, must be proportional to $1/r^2$.

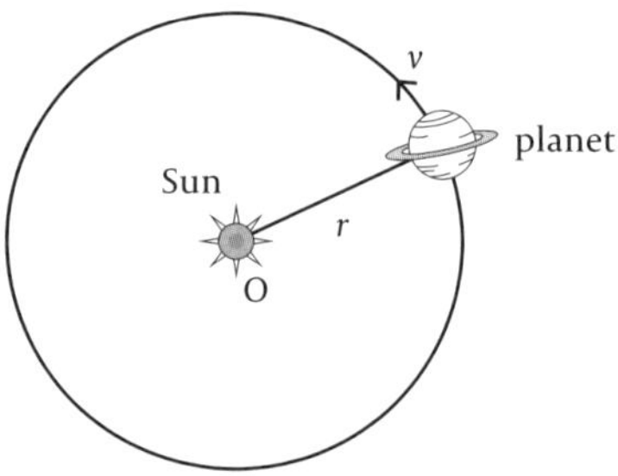

Fig. 35 The circular approximation to planetary motion.

And as acceleration is caused by a force, this suggests that the planets experience an attractive force towards the Sun which is *proportional to* $1/r^2$.

* * * * * * *

By 1679, the whole question of planetary motion had become the 'hot topic' of the day, as is again evident from Hooke's diary:

> *October 18th* Gresham College about Elliptick motion.
> *October 21st* At Bruins coffee house with Sir Chr. Wren about . . . coyled cone for Celestiall theory.

The passing reference to a 'coyled cone' reveals the way in which Hooke resorted, once again, to mechanical devices to loosely simulate planetary motion, when what was really needed was groundbreaking mathematical theory.

In any event, on 24 November 1679, Hooke wrote to Newton, asking for any thoughts about his idea of

> compounding the celestiall motions of the planetts of a direct motion by the tangent and an attractive motion towards the centrall body.

In his reply, Newton claimed to be completely unaware of Hooke's idea, yet the letter seems to have helped him to clarify his own thoughts on the whole matter. And he soon took up the problem afresh, albeit with his legendary secrecy.

For it was not until five years later, in 1684, that Edmund Halley made his famous journey to Cambridge in order to ask Newton the key question at the time: what would the orbit of a planet be if it were attracted to the Sun by a force obeying an *inverse-square* law?

According to a contemporary:

> Sr Isaac replied immediately that it would be an Ellipsis, the Doctor struck with joy & amazement asked him how he knew it, why saith he I have calculated it

And it was in this way, somewhat unwittingly, that Halley's visit led, three years later still, to Newton's great masterpiece—the *Principia* . . .

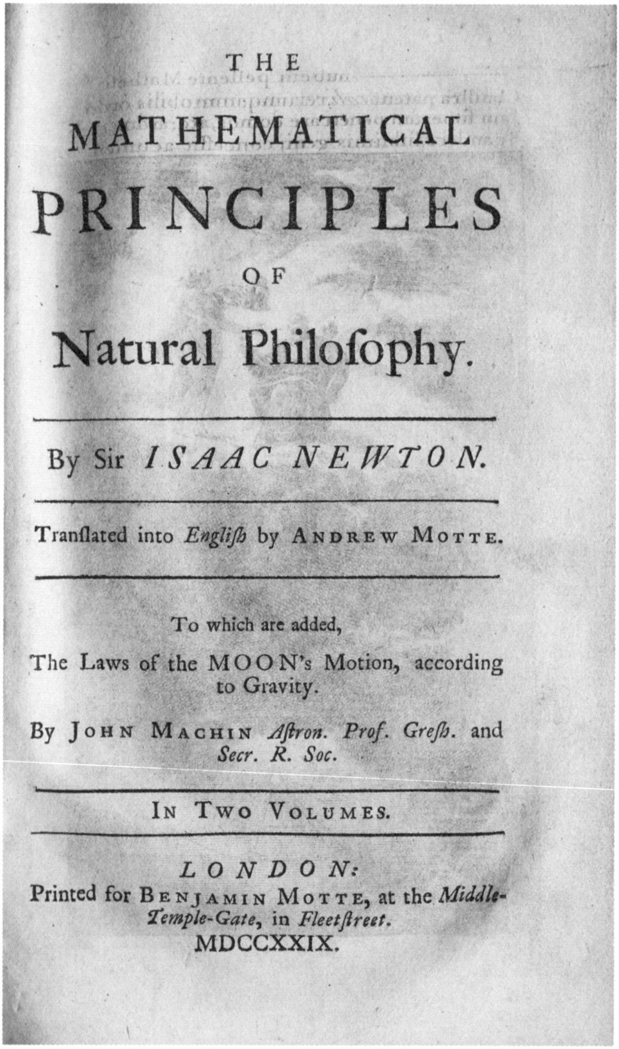

THE

MATHEMATICAL

PRINCIPLES

OF

Natural Philoſophy.

By Sir *ISAAC NEWTON.*

Tranſlated into *Engliſh* by ANDREW MOTTE.

To which are added,

The Laws of the MOON's Motion, according to Gravity.

By JOHN MACHIN *Aſtron. Prof. Greſh.* and *Secr. R. Soc.*

IN TWO VOLUMES.

LONDON:

Printed for BENJAMIN MOTTE, at the *Middle-Temple-Gate*, in *Fleetſtreet.*

MDCCXXIX.

Fig. 36 The first version in English of Newton's *Principia* (1729).

9

PRINCIPIA

Opening Newton's *Principia* for the first time can be something of a shock.

The most obvious example is perhaps Newton's own statement of his famous second law of motion, which reads as follows:

> The alteration of motion is ever proportional to the motive force impress'd; and is made in the direction of the right line in which that force is impress'd.

Now, Newton means by 'motion' what we now call *momentum*, and the usual modern interpretation of his statement is then

$$\textit{force} = \textit{rate of change of momentum}.$$

And when the force in question acts on a single and *constant* point mass m, this does indeed turn out to be entirely equivalent to

$$\textit{force} = \textit{mass} \times \textit{acceleration}.$$

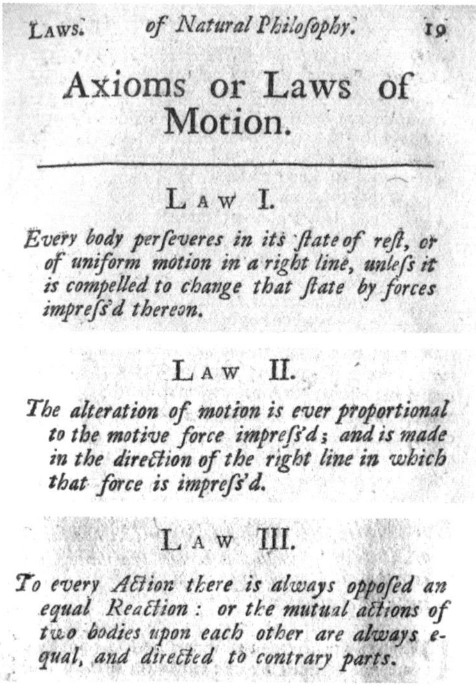

LAWS. *of Natural Philoſophy.* 19

Axioms or Laws of Motion.

LAW I.

Every body perſeveres in its ſtate of reſt, or of uniform motion in a right line, unleſs it is compelled to change that ſtate by forces impreſs'd thereon.

LAW II.

The alteration of motion is ever proportional to the motive force impreſs'd; and is made in the direction of the right line in which that force is impreſs'd.

LAW III.

To every Action there is always oppoſed an equal Reaction: or the mutual actions of two bodies upon each other are always equal, and directed to contrary parts.

Fig. 37 Newton's three laws of motion, in the first English edition of his *Principia* (1729). The Latin original was published in 1687.

There is, however, an altogether different way of viewing Newton's own statement of his second law, in which 'alteration' is simply 'change', rather than 'rate of change'. And on the assumption that the change in question is meant to be *sudden*, Newton's 'force' is then not what we would call a force at all, but what is today called an 'impulse', namely a large force acting for a very short time, which does indeed produce a more or less instantaneous change in momentum.

And, as it happens, this second interpretation accords well with how Newton actually tackles many problems in the *Principia*.

* * * * * * *

It also accords well with the important role played by *collisions* in the discovery of the laws of motion, and Newton did, in fact, conduct his own experiments on the subject as early as 1664, even considering the forces involved.

For, as the total momentum in a collision is conserved, the two bodies involved—which Newton calls *p* and *r*—must exchange momentum during the collision itself in equal amounts. And in his *Waste Book* of 1664 Newton presents an explanation of this:

> for soe much as *p* presseth upon *r*, so much *r* presseth on *p*.

In this short passage, then, we have an early indication of both the link between force and change of momentum *and* Newton's *third* law of motion in the *Principia*, usually paraphrased today as

> action and reaction are equal and opposite.

(see Fig. 37).

* * * * * * *

Yet there is a second and quite different surprise that lies in wait for any reader new to the *Principia*, and this is the sheer amount of *geometry*.

It is simply everywhere.

Newton's interest in the subject appears to have begun in the autumn of 1663, when he visited a fair in Cambridge and purchased a book on astrology. For this led him to trigonometry and so, eventually, to Euclid's *Elements*.

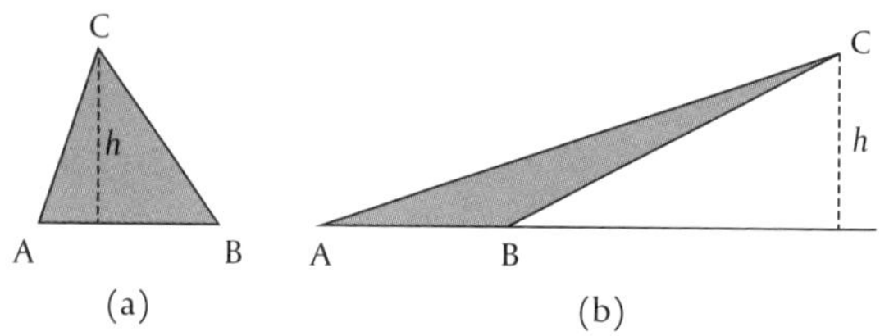

Fig. 38 Same area! From Book 1, Proposition 38 of Euclid's *Elements*.

While Newton eventually took Euclid's great masterpiece as a model for the *Principia*, he seems at first to have been somewhat underwhelmed by it, finding the early propositions so intuitively obvious that he

> wondered how anybody would amuse themselves to write any demonstrations of them.

But one particular result, it seems, really surprised him and caused him to change his mind: the two triangles in Fig. 38 have *the same area*, because they have equal bases AB and the same 'height' *h*. And they will continue to have the same area, no matter how far we move the point C—provided we move it always parallel to AB.

And, whether by coincidence or not, this particular idea played a key role in one of the greatest results in the *Principia* ...

10

NEWTON AND PLANETARY MOTION

We have seen already that the key problem at the time was to account for planetary motion, if possible, by a gravitational force directed always towards the Sun.

And one of the very earliest results in Newton's *Principia* shows that Kepler's equal-area law can be explained in just this way.

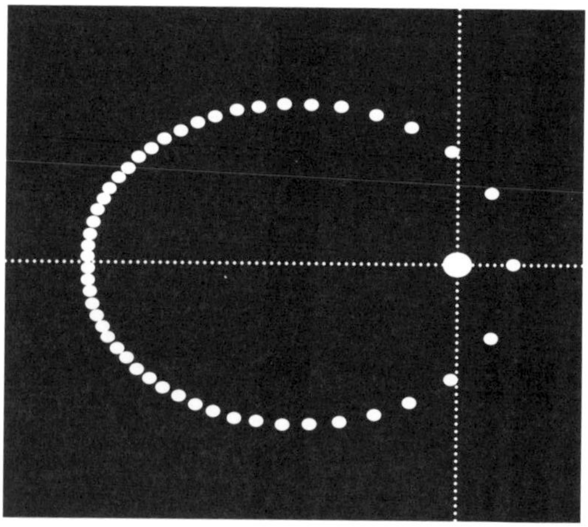

Fig. 39 Another view of planetary motion.

Motion under a central force

To do this, Newton replaces a continuously acting force with a whole sequence of short sharp *impulses* towards the fixed point, S (Fig. 40).

Now, in the absence of any such impulse, the planet, having moved at constant velocity from A to B, would continue in a straight line at that same velocity, and after an equal time interval would have reached C, where BC = AB. And by the geometrical result in Fig. 38, the triangular areas SAB and SBC will be equal.

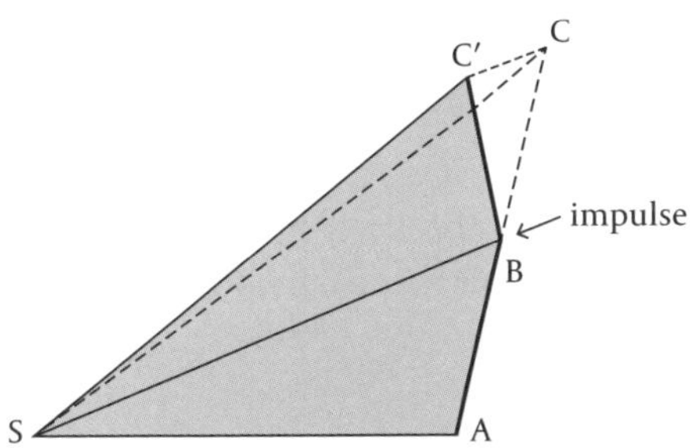

Fig. 40 Newton's approach to the problem of motion under a central force.

Now, an impulse at the point B, acting towards S, produces instead an instantaneous change in velocity in such a way that the planet ends up at C′ instead of C, where CC′ is proportional in magnitude to the impulse and *parallel to* SB. By the same geometrical result, the areas SBC′ and SBC must therefore be equal.

It then follows immediately that the areas SAB and SBC′ must be equal, and it is in this way that Newton explains how Kepler's equal-area law follows from the assumption of a central force on each planet directed always towards the Sun.

While this was an enormous leap forward, it is important to note that this sweeping out of equal areas in equal times holds for *any* force that acts always towards the fixed point S.

In other words, and in the context of planetary motion, it tells us nothing, in itself, about how the gravitational force on a planet depends on the distance *r* from the Sun.

Finding the law of force

To establish this, Newton imagines the planet moving from the point P in Fig. 40 to the nearby point Q a very short time later.

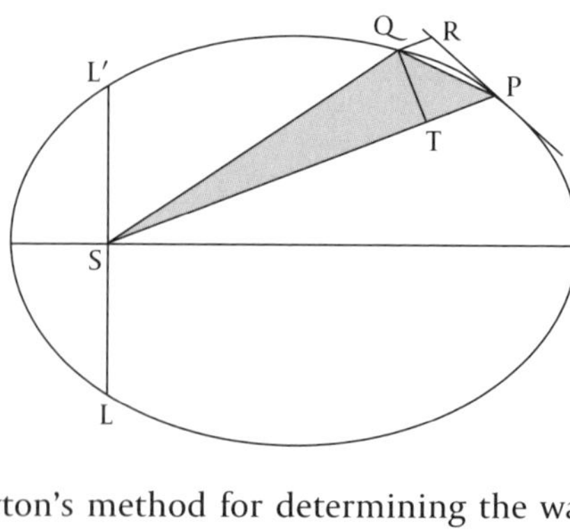

Fig. 41 Newton's method for determining the way in which a central force depends on the distance SP.

He draws the tangent at P, then the line QR parallel to SP, and finally QT perpendicular to SP.

We may then paraphrase Newton's argument as follows. First, the time interval between P and Q is very small, so the force towards S will be almost constant during that time interval, leading to an acceleration a towards S that is almost constant. From our Chapter 2, then, QR will be approximately $\frac{1}{2}at^2$, where t is the elapsed time between P and Q. And as the force will be proportional to the acceleration a, it will be proportional to QR/t^2.

Newton now eliminates time, using the equal-area law. For the area swept out, SPQ, will be proportional to the elapsed time t, and that area will, in turn, be almost exactly the area of the triangle SPQ, which is $\frac{1}{2}\,\mathrm{QT} \times \mathrm{SP}$.

This means that the force towards S must be proportional to

$$\frac{\mathrm{QR}}{(\mathrm{QT})^2} \times \frac{1}{(\mathrm{SP})^2}.$$

* * * * * * *

Now, up until this point the argument has been entirely general, valid for *any* curve. In the final step, however, Newton uses some fairly obscure theorems concerning the geometry of the ellipse to show that as we take Q closer and closer to P in Fig. 41 the ratio $\mathrm{QR}/(\mathrm{QT})^2$ gets closer and closer to $1/(\mathrm{LL'})$, which is a *constant*.

In this way, Newton shows that if the orbit is an ellipse, with the force directed always to one focus, then at distance r from that focus,

$$\textit{force towards } S = \frac{\textit{constant}}{r^2}.$$

And from the point of view of the history of dynamics, this is, arguably, the greatest single moment in the whole of the *Principia*.

11

CALCULUS

Newton's *Principia* of 1687 is now acclaimed as one of the greatest masterpieces in mathematics and science. At the time, however, it had its problems.

One difficulty was that very few people could really understand it properly, let alone build upon it and take the subject further. Even the great Swiss mathematician Leonhard Euler wrote in 1736:

> When I began to examine Newton's *Principia* . . . even though I thought I could understand the solutions . . . I could not solve problems that were slightly different.

The difficulty lay not simply in Newton's highly geometrical methods, but in the way that so many problems seemed to require some fresh new approach, peculiar to the particular example under consideration.

In the event, it took about 60 years from Newton's masterpiece for continental mathematicians to slowly fashion classical mechanics into more or less the form we know today.

And the key innovation was *calculus*—a subject due, very largely, to Newton himself and to Gottfried Wilhelm Leibniz in Germany.

* * * * * * *

The most central idea of calculus is the *rate* at which things change.

Suppose, for instance, that some physical quantity y is changing with time t. Then, in a notation due to Leibniz, the *rate* at which y is changing is denoted by the slightly curious but highly effective symbol dy/dt (Fig. 42).

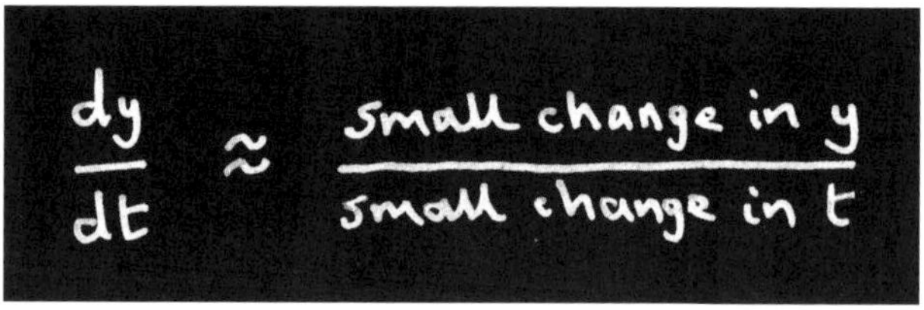

Fig. 42 An informal view of the rate at which y changes with time t.

We may, if we wish, think of dy/dt loosely as just a small change in y divided by the corresponding small change in t. We may, in other words, regard d here as just shorthand for the phrase 'a small change in . . .'.

The full picture, however, is rather more subtle. For what dy/dt *really* means is the value approached by the ratio in Fig. 42 *as both changes get smaller and smaller.*

This whole, rather subtle process is called *differentiation*, and a perfectly valid alternative way of thinking about the whole idea is simply to view *d*/*dt itself* as a symbol, meaning

$$\frac{d}{dt} = \textit{rate of change.}$$

* * * * * * *

A good example is provided once again by the (somewhat hypothetical) object falling from the Leaning Tower of Pisa (Fig. 43).

Fig. 43 The Leaning Tower experiment revisited.

We saw in Chapter 2 that the distance *s* fallen in time *t* is

$$s = \frac{1}{2}gt^2.$$

How, then, can we deduce from this formula the *speed* at which the object is falling at any given moment?

Well, we will presumably get a good approximation to the speed at time t by using the formula itself to calculate the short *extra* distance fallen after a short *extra* time, and then dividing one by the other. Moreover, the shorter we make the 'extra' time, the closer we should presumably get to what we really want—the speed actually at time t.

This all amounts to precisely the 'differentiation' procedure just described, as indicated in Fig. 43, and if we actually carry it out, to find ds/dt, we find that the downward speed v is

$$v = gt$$

at time t (Notes, p. 233).

In this way, then, it is possible to confirm some of the results concerning constant acceleration in Chapter 2, which we obtained there the other way round, by starting with the speed formula and then deducing the one for the distance travelled.

* * * * * * *

Calculus is more or less everywhere in dynamics, often because of the fundamental law

$$\textit{force} = \textit{mass} \times \textit{acceleration},$$

and the fact that acceleration is rate of change of velocity.

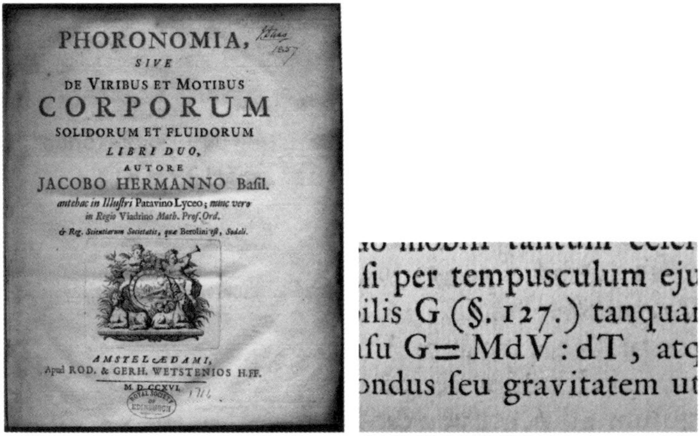

PHORONOMIA,
SIVE
DE VIRIBUS ET MOTIBUS
CORPORUM
SOLIDORUM ET FLUIDORUM
LIBRI DUO,
AUTORE
JACOBO HERMANNO Basil.
antehac in Illustri Patavino Lyceo; nunc vero
in Regio Viadrino Math. Prof. Ord.
& Reg. Scientiarum Societatis, quæ Berolini est, Sodali.
AMSTELÆDAMI,
Apud ROD. & GERH. WETSTENIOS H.FF.
M.D.CCXVI.

ſi per tempusculum eju
ilis G (§. 127.) tanqua
ſu G = MdV : dT, atq
ondus ſeu gravitatem ut

Fig. 44 The first appearance of Newton's second law of motion in calculus form, in Jacob Hermann's *Phoronomia* of 1716.

Suppose, for instance, that a mass m is moving in a straight line, and that we know how its velocity v depends on time t. Then the force F on it will be given by

$$F = m\frac{dv}{dt}.$$

And this would give a fair indication of how much of modern dynamics is done today—except for one thing . . .

* * * * * * *

More often than not, the real problem in dynamics is precisely *the other way round.*

We typically start, in other words, by knowing something about the forces involved, and the fundamental problem is then to determine how the system itself evolves with time.

In the equation above, for example, we would normally start by knowing something about the force F, and therefore something about the rate at which the velocity v is changing. The equation itself then becomes a so-called *differential equation*, to be solved for v in terms of time t.

All too often, in other words, the main problem in dynamics is to *undo* or *reverse* the whole differentiation process, and this quickly turns out to be much more challenging—and intriguing—than differentiation itself.

12

OSCILLATIONS

A good example of the new calculus methods in action is provided by a toy spider that hangs on a long spring at my home in Oxford (Fig. 45).

It is at least 60 years old, with only six legs and a replacement nose, yet, when displaced, it still bobs happily up and down.

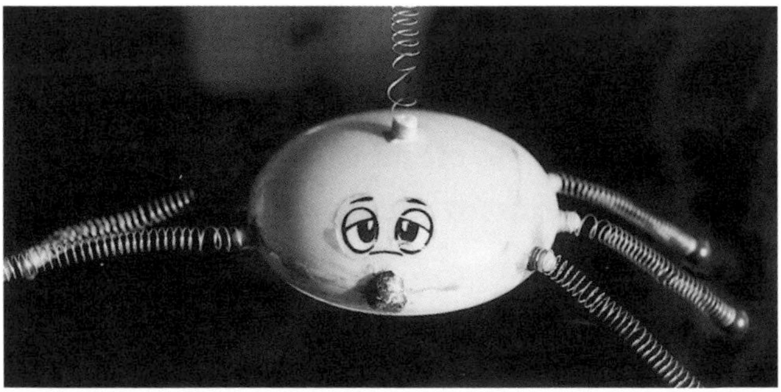

Fig. 45 A toy spider on a spring.

Now, Hooke's law of elasticity says that the restoring force exerted by a spring is proportional to its extension (or compression). So, if k denotes the constant of proportionality, the spider will hang at rest with extension l such that

$$kl = mg,$$

so that the upward force due to the spring exactly balances the downward force due to gravity.

* * * * * * *

To investigate the spider's *motion*, let y denote its upward displacement from the natural hanging position, so that dy/dt denotes its upward velocity.

The upward *acceleration* will therefore be the rate of change of dy/dt, and the conventional notation for this, again due to Leibniz, is

$$\frac{d^2y}{dt^2} = \frac{d}{dt}\left(\frac{dy}{dt}\right).$$

The net upward force on the spider, which will be solely due to the spring, is $-ky$, so the fundamental law *force* = *mass* × *acceleration* then leads to the differential equation

$$\frac{d^2y}{dt^2} = -\omega^2 y,$$

where $\omega^2 = k/m$.

Our problem, then, is to solve this equation, subject to some *initial conditions*, to find how y depends on time t.

* * * * * * *

Now, the most obvious way of starting the whole motion is to give the spider an upward displacement a, say, and release it from a state of rest. The initial conditions are then $y = a$ and $dy/dt = 0$ when $t = 0$.

Calculus methods can then be used to solve the problem, though the solution is not remotely 'obvious':

$$y = a \cos \omega t$$

(Notes, p. 234).

And, as Fig. 46 shows, cos ωt *oscillates*, in a particular manner known as *simple harmonic motion*, which is, in fact, a characteristic feature of many other oscillating systems, as we will see later in the book.

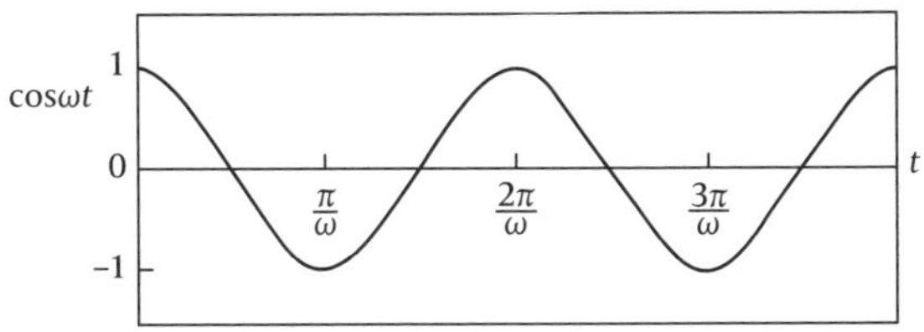

Fig. 46 The graph of cos ωt.

Most importantly, the time for one complete oscillation, known as the oscillation *period*, is evidently given by

$$T = 2\pi/\omega.$$

An experiment

In the case of the spider, then, the time for each complete oscillation will be $T = 2\pi\sqrt{m/k}$, which is independent of the initial displacement a, and determined solely by the mass of the spider and the strength of the spring.

And, plainly, a stronger spring, with larger k, results—as one would expect—in a smaller value of T and a more rapid oscillation.

I decided to test this particular result by experiment, but I had no obvious practical way of measuring directly either the spider's mass m or the spring strength k.

Fortunately, the relation $kl = mg$, noted earlier, allowed me to determine the ratio k/m simply by measuring the extension l when the spring was hanging at rest, and using the fact that $g = 9.81 \text{ m s}^{-2}$.

Even this was a bit tricky, as the spring itself was so old and rickety that its natural or 'unstretched' length wasn't terribly clear, and two different approaches gave the extension l as 32 or 34 cm. On taking the average, the oscillation period T worked out as 1.15 seconds, corresponding to 52 oscillations per minute.

When I did the experiment, I counted 50, rather than 52, but I was still very happy indeed, in the circumstances, with an error of just 4 per cent.

13

A NEW AGE BEGINS...

If we ask where classical mechanics was first done in this new kind of way, the answer is to be found in various papers, mostly by Euler, in the period 1730–50.

Leonhard Euler was born in Basel, Switzerland in 1707, and studied with Johann Bernoulli. In 1727 he took up an appointment in St Petersburg, working alongside Johann's son Daniel, and while there he published *Mechanica* (1736),

(a) (b)

Fig. 47 (a) Leonhard Euler. (b) Euler's signature and date on a letter of 1754 in the author's possession.

a groundbreaking application of calculus methods to mechanics.

In 1741, he moved to the Academy of Sciences in Berlin, where one academic visitor observed:

> Leonhard Euler is not, like the great algebraists usually are, of sinister character and clumsy behaviour, but cheerful and lively.

And it was during Euler's 25 years in Berlin that he made many of his greatest contributions to mechanics.

In particular, it is in a paper of 1749 that we find, for the very first time, the general *differential equations* for a point mass *m*, set up much as we would today, with components of both force and acceleration taken with respect to a set of rectangular *coordinates* (x, y, z).

Cela posé, prenant l'element du tems dt pour conſtant, le changement inſtantané du mouvement du Corps ſera exprimé par ces trois équations :

$$\text{I. } \frac{2ddx}{dt^2} = \frac{X}{M}; \quad \text{II. } \frac{2ddy}{dt^2} = \frac{Y}{M}; \quad \text{III. } \frac{2ddz}{dt^2} = \frac{Z}{M}$$

d'où l'on pourra tirer pour chaque tems ecoulé t les valeurs x, y, z, & par conſéquent l'endroit où le Corps ſe trouvera. C. Q. F. T.

Fig. 48 The first general differential equations of motion for a point mass, by Euler, in 1749. His components of force are X, Y, Z, and the unexpected factor of 2 is explained in the Notes, p. 234.

Today, for instance, we might write the first of the three equations in Fig. 48 in the form

$$m\frac{d^2x}{dt^2} = F_x,$$

where F_x denotes the x-component of force.

Polar coordinates

As it happens, Euler switches quite rapidly in his paper of 1749 to motion in just two dimensions and the so-called *polar* coordinates of Fig. 49.

Here, the position of the point mass P is specified by the distance r from some fixed point O and the angle θ (measured in radians) which OP makes with some fixed line through O.

The velocity components in the 'radial' and 'transverse' directions then turn out to be

$$v_r = \frac{dr}{dt}, \quad v_\theta = r\frac{d\theta}{dt}.$$

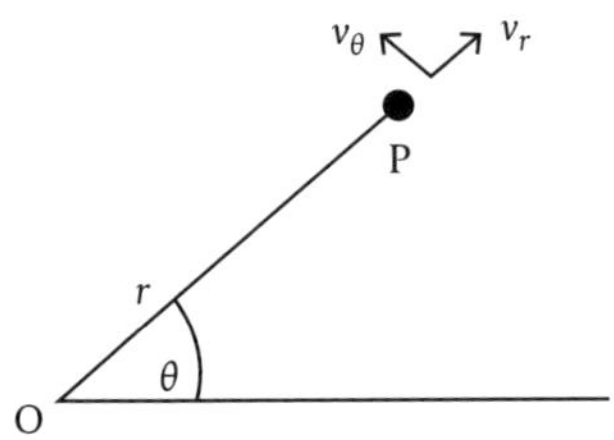

Fig. 49 Polar coordinates.

And if we denote the components of force in the same way—with F_r in the direction of OP, and F_θ perpendicular to it—and apply the law *force* = *mass* × *acceleration*, we find from the radial component that

$$m\left(\frac{dv_r}{dt} - \frac{v_\theta^2}{r}\right) = F_r.$$

This is, arguably, no great surprise. In the special case of circular motion about O, for instance, so that $v_r = 0$, it tells us that a radially *inward* force mv_θ^2/r is needed to maintain the motion, which we found by a quite different argument in Chapter 4.

When we apply *force* = *mass* × *acceleration* in the *transverse* direction, however, perpendicular to OP, something rather more unexpected happens:

$$m\frac{d}{dt}(rv_\theta) = rF_\theta,$$

and we will see later in the book how this is, in fact, one special case of a new—and, more usually, independent—law of mechanics.

Central force motion

For present purposes, however, the most significant feature of all this is that if the force is *central*, i.e. directed always towards O, then $F_\theta = 0$, and rv_θ is therefore constant.

And as $v_\theta = rd\theta/dt$ this leads in turn to the major result in Fig. 50.

At this point it is convenient to introduce the idea of the *angular velocity* of P about O, which is just the rate at which the line OP is rotating:

$$\textit{angular velocity} = d\theta/dt,$$

though we should note that this is not a 'velocity' in the usual sense at all, for its dimensions are not L/T but 1/T, the angle θ being a pure number and dimensionless.

According to Fig. 50, then, if r increases during motion under a central force, the angular velocity $d\theta/dt$ must decrease, and vice versa.

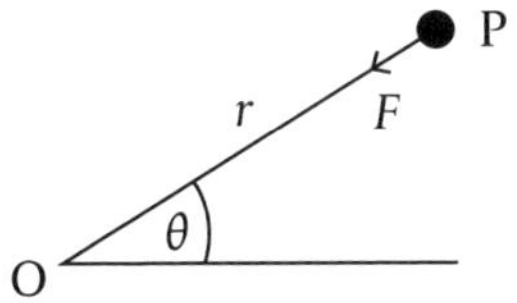

$$r^2 \frac{d\theta}{dt} = \text{constant}$$

Fig. 50 Motion under a central force.

And while it is far from obvious, this same result also implies that the line OP *sweeps out area at a constant rate* (Notes, p. 234).

It is in this way, then, that Euler's new calculus-based approach to dynamics accounts for one of Kepler's laws of planetary motion.

And, as it happens, the result in Fig. 50 also provides the key to a rather more fanciful problem . . .

14

‘CONTRARY TO Y^E OPINION OF Y^E VULGAR’

Imagine, if you will, dropping an object from a high tower on the Earth’s equator.

Bearing in mind the rotation of the Earth itself, will the object land:

(a) slightly to the west,
(b) slightly to the east, or
(c) exactly at the bottom of the tower?

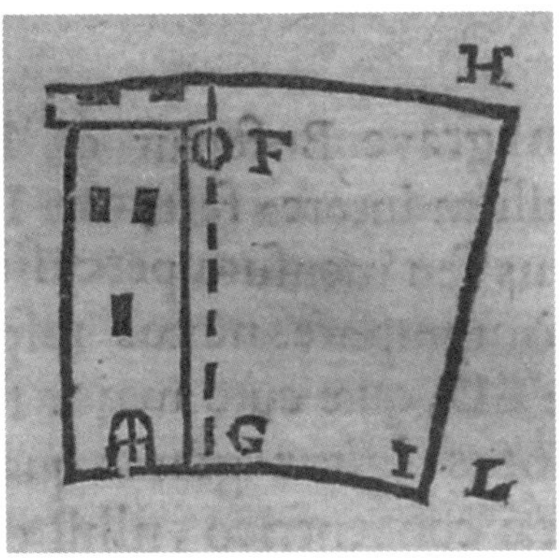

Fig. 51 From an early account of the ‘high tower’ problem, in Dechales’ *Cursus seu Mundus Mathematicus* (1674).

As it happens, Newton posed just this problem when he replied to Hooke's thought-provoking letter of 1679 about planetary motion.

He claims that common opinion would have the object falling slightly to the *west*, on the grounds that, during its fall, the Earth itself would have turned slightly to the east.

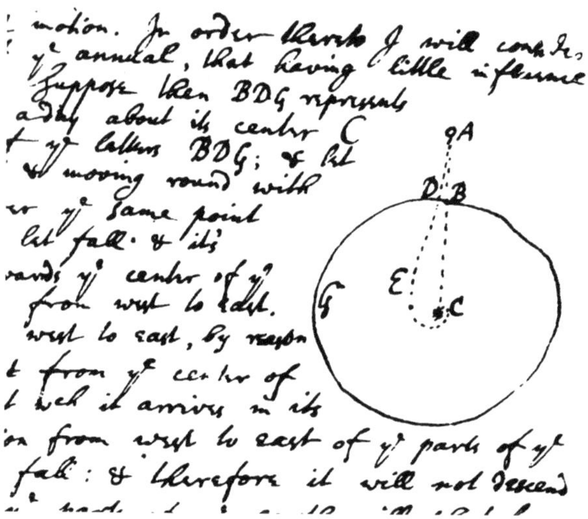

Fig. 52 Newton's own sketch of an object falling from a high tower, relative to the Earth itself, and viewed from some point above the North Pole. (His imagined continuation of the path—somehow—*into the Earth itself* is another story.)

But such an argument completely ignores, of course, the fact that the object is initially *rotating with the Earth*. And I imagine that a more common opinion today might therefore be that the object will land exactly at the bottom of the tower.

Yet Newton argues, instead, that the object

> outrunning ye parts of ye earth will shoot forward to ye east side of the perpendicular ... quite contrary to ye opinion of ye vulgar.

So, *who is right?*

* * * * * * *

Well, this is, in fact, a central-force problem, because the gravitational pull of the Earth is towards its centre.

And we know from Chapter 13 that central-force motion is characterized by the important result in Fig. 50, which turns out to be just what we need for the tower problem.

For the object and the Earth will initially have the *same* angular velocity $d\theta/dt$. But as the object falls, r decreases slightly, leading to an *increase* in its angular velocity, so that the object does indeed

> shoot forward to y^e east side of the perpendicular

just as Newton claimed in his letter to Hooke.

It seems to me doubtful, however, whether Newton's reasoning at the time can possibly have been correct, because—so far as I can determine—he didn't have either the result in Fig. 50 *or* (at that time) its equivalent in terms of constant-sweeping-of-area (see Notes, p. 235).

A modern calculation shows the actual amount of eastward displacement to be

$$\frac{2}{3}\Omega\sqrt{\frac{2h^3}{g}},$$

where Ω is the angular velocity of the Earth, and h the height of the tower.

The effect is small, but not absurdly so. According to my calculations, if the Eiffel tower were on the equator the deflection to the east due to the Earth's rotation would be about 11 cm.

15

THE PRINCIPLE OF LINEAR MOMENTUM

Classical mechanics extends well beyond the motion of single point masses, to rigid bodies, fluids, and elastic solids.

And there are two fundamental laws on which much of the theory is based, both due largely to Euler.

DECOUVERTE
D'UN NOUVEAU PRINCIPE DE MECANIQUE,
PAR M. EULER.

I.

Un corps eſt appellé ſolide, dont l'intérieur n'eſt aſſujetti à aucun changement, ou dont toutes les parties conſervent conſtamment la même ſituation entr'elles, quelque mouvement que puiſſe avoir le corps entier. Nonobſtant cette ferme & invariable liaiſon des parties, un corps ſolide peut recevoir une infinité de mouvemens differens, dont la determination & les loix, qui s'y obſervent, ſont

Fig. 53 Euler's paper of 1752 announcing the principle of linear momentum.

The first is *the principle of linear momentum* (1752):

> The total external force on a body is equal to the rate of change of its momentum.

At first sight, perhaps, this may appear to be just a slight paraphrase of Newton's second law, but Euler himself certainly didn't see it like that, not least because the title of his 1752 paper was 'The Discovery of a New Principle of Mechanics' (Fig. 53).

For it is in this paper that we find it stated clearly for the first time that the 'body' in question may be a single point mass, a whole system of such point masses, or *each and every part* of a continuous medium such as a rigid body or fluid.

Consider, for instance, the fluid in Fig. 54, and suppose for simplicity that the only forces acting are the internal pressure forces between the various bits of fluid. And let us focus on one particular (darkly shaded) blob within the fluid, which we may view as consisting of two contiguous parts, A and B.

Now, we are certainly entitled to apply the principle of linear momentum to the shaded blob as a whole, in which case the only external forces are due to the part C. Yet we are also entitled to apply the same principle to the parts A and B separately, but when doing so must then include the forces that each exerts on the other.

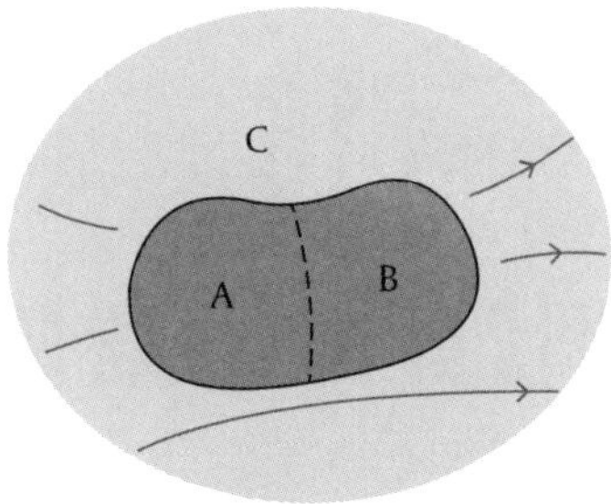

Fig. 54 Applying the principle of linear momentum.

And it is relatively easy to show that the three results are then consistent with one another only if the forces that A and B exert *on each other* are equal and opposite (Notes, p. 235).

In this way, then, Euler's principle of linear momentum builds not only on Newton's second law, but on his third law as well.

* * * * * * *

Yet it turns out that the principle of linear momentum *alone* is not, in general, sufficient as a basis for mechanics, and during the eighteenth century it gradually emerged that a further principle was needed, concerning *rotational* motion . . .

16

THE PRINCIPLE OF ANGULAR MOMENTUM

One of the most iconic examples of rotational dynamics is the ice-skater of Fig. 55.

At first, she spins slowly, with her arms outstretched, but she then brings her arms in, and, in consequence, spins more rapidly.

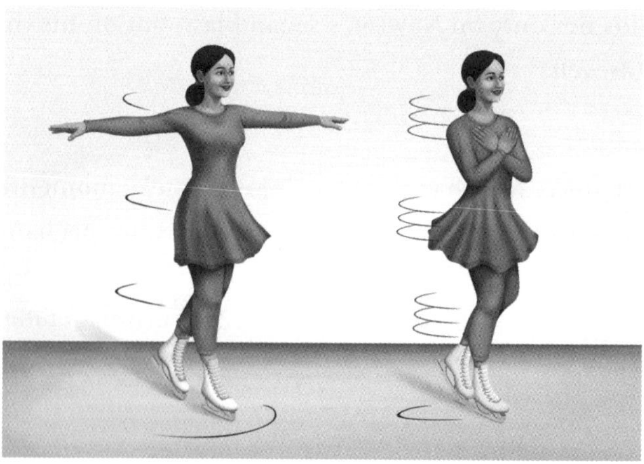

Fig. 55 A skater spinning on ice.

And to understand why, we need to introduce two new concepts.

Torque

Loosely speaking, a torque reflects the capacity of a force to *turn* something about some point O, and it is, in fact, a very old concept in mechanics, dating back to Archimedes' law of the lever.

In the example of Fig. 56a, for instance, the *torque* about O is the product rF, obtained by taking the force component F perpendicular to OP and multiplying by the distance r. (This is also known as the *moment* of the force F about O.)

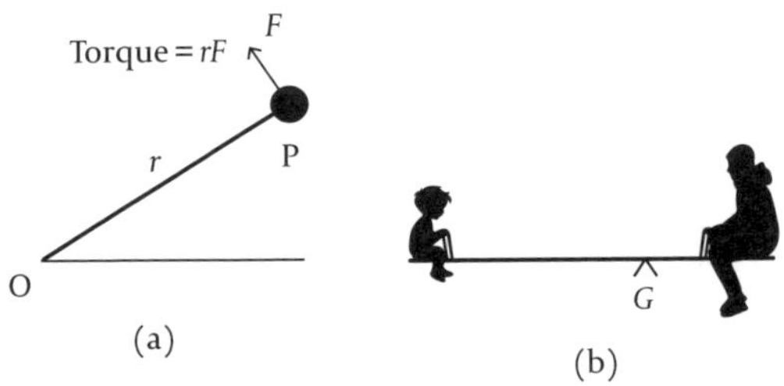

Fig. 56 (a) The torque about O. (b) Archimedes' law of the lever in action.

Figure 56b shows an example of the whole idea in action, where a child on a see-saw balances someone much heavier by sitting further from the pivot, in just such a way that the two torques cancel out.

And the point G about which the system balances as a whole is called the *centre of gravity*.

Angular momentum

This is perhaps less intuitive as an idea, but it is nevertheless defined for the point mass at P in Fig. 57 in exactly the same way—we take the component of momentum perpendicular to OP, namely mv, and multiply by r to get rmv.

As with angular velocity, the terminology, though standard, is not great, because angular momentum isn't a *momentum* in the usual sense at all; it plainly has dimensions of momentum × length.

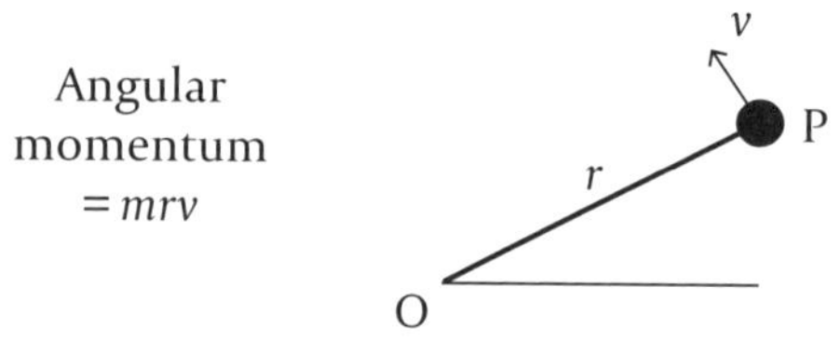

Fig. 57 The angular momentum (or 'moment of momentum') about O of a point mass m, where v denotes the transverse component of velocity.

* * * * * * *

To explore the whole idea further it is best, I think, to consider a rigid body rotating about a fixed axis. For its angular momentum is then proportional to its *angular velocity* ω about that axis, which, with reference to Fig. 58, is defined as

$$\omega = d\theta/dt,$$

where θ denotes the angle through which the body rotates in time t.

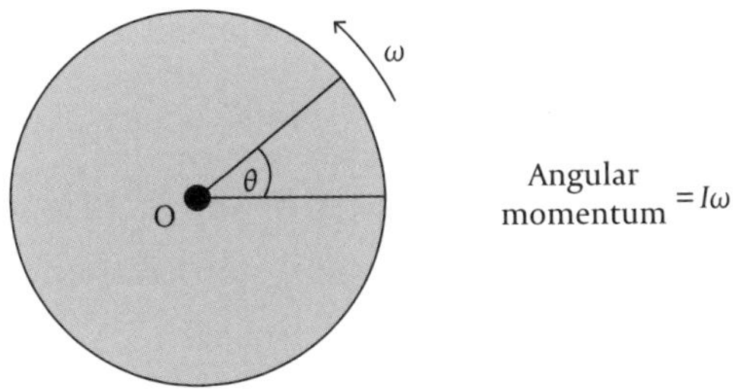

Fig. 58 Angular momentum of a rigid body.

The constant of proportionality I is called the object's *moment of inertia* about that axis, and this depends on both the total mass M and the way in which that mass is distributed. In particular, the moment of inertia I is greater when the mass is concentrated further from the rotation axis, making the object harder to turn.

If, for instance, all the mass in Fig. 58 were to be concentrated on a circle of radius a, a bit like a bicycle wheel, the moment of inertia would be Ma^2, but if the mass were to be spread out over a uniform disc of the same radius the moment of inertia would be only $\frac{1}{2}Ma^2$.

The principle of angular momentum

The two ideas of torque and angular momentum pave the way for a further major principle of mechanics, once again due largely to Euler—*the principle of angular momentum*:

> The total torque on a body about some fixed point O is equal to the rate of change of the angular momentum about O.

Moreover, the same principle can be shown to work equally well *about the centre of gravity* of a system, rather than about a fixed point.

This explains, for instance, what happens to the ice-skater in Fig. 55. For with essentially no torque about a vertical axis from contact with the ice, her angular momentum $I\omega$ must stay *constant*. So, when she brings her arms in, reducing her moment of inertia I, her angular velocity ω has to increase as a result.

And if we were to apply a torque Γ, say, to the object in Fig. 58, then, because the moment of inertia I is constant in that case, the principle of angular momentum tells us that

$$\Gamma = I\frac{d\omega}{dt},$$

so that the torque produces, as one would expect, an increase in angular velocity.

* * * * * * *

In practice, the way in which the principle of angular momentum is viewed and used today depends on the context.

In the case of a single point mass, for instance, it can be deduced directly from the principle of linear momentum, or indeed from Newton's second law, as we saw in Chapter 13. And the same is true for a whole system of point masses which attract one another by gravitational forces. Another such example is a nonviscous fluid.

Yet with viscous fluids and, especially, elastic solids, the situation is very different, and the principles of linear and angular momentum tend to be seen as essentially independent, and equally fundamental laws of mechanics (Notes, pp. 236–8).

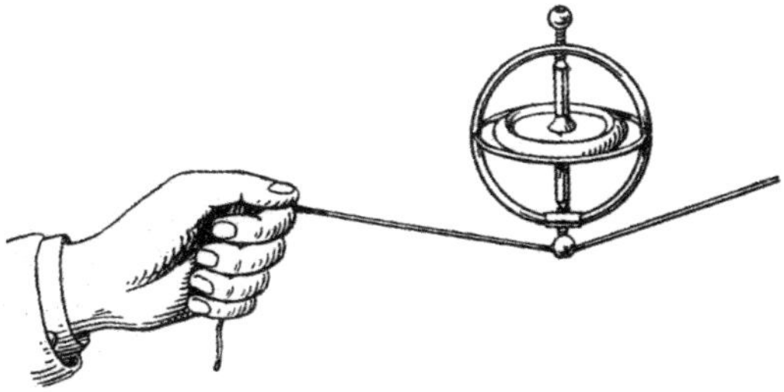

Fig. 59 Rigid body mechanics in action, from Crabtree's *Spinning Tops and Gyroscopic Motion* (1909).

Finally, there is the case of rigid body motion, which is a little odd in this particular respect. This is because the vast majority of twentieth-century textbooks on the subject claim to *deduce* the principle of angular momentum for a rigid body by taking moments and using Newton's second law. To do this, however, they typically make an assumption about internal forces that is arguably spurious and even a little bizarre, and most certainly not what Euler actually did (Notes, pp. 237–8).

Yet it cannot be denied that rigid body dynamics is an excellent way of seeing the principle of angular momentum actually at work, and that is, in fact, where we turn next.

17

SPINNING TOPS

The photograph in Fig. 60 shows two of the greatest quantum physicists of the twentieth century, Wolfgang Pauli and Niels Bohr, looking down at a tiny spinning top which is doing something very strange on the floor.

Fig. 60 Pauli and Bohr enjoying classical mechanics at the Institute of Physics in Lund, Sweden, 1954.

For this is no ordinary spinning top. If spun fast enough, with its centre of gravity in its lowest position, it becomes unstable, and slowly turns over until its stem eventually touches the ground. It then scoots about, rises on the stem, and eventually turns completely *upside-down* (Fig. 61).

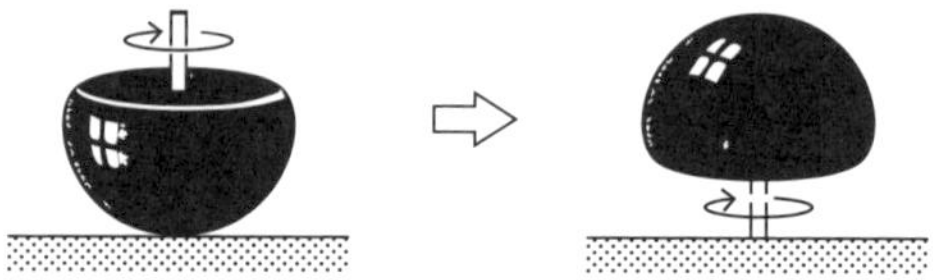

Fig. 61 The tippy-top in action.

* * * * * * *

Some might say that a normal spinning top is peculiar enough, in view of the way that rapid rotation somehow stops it from falling over. And one of the most curious aspects of such a top is the way that its axis of symmetry slowly *precesses* about the vertical.

The dotted line in Fig. 62 shows the phenomenon for a top consisting of a uniform circular disc of radius a mounted on a weightless axle, at a distance h from the pivot O.

In practice, the kinetic energy of the top is usually much greater than its gravitational potential energy, and it turns out that the angular velocity of precession is then given by

$$\Omega \approx \frac{2gh}{a^2\omega},$$

where ω is the angular velocity of spin of the top about its own axis of symmetry.

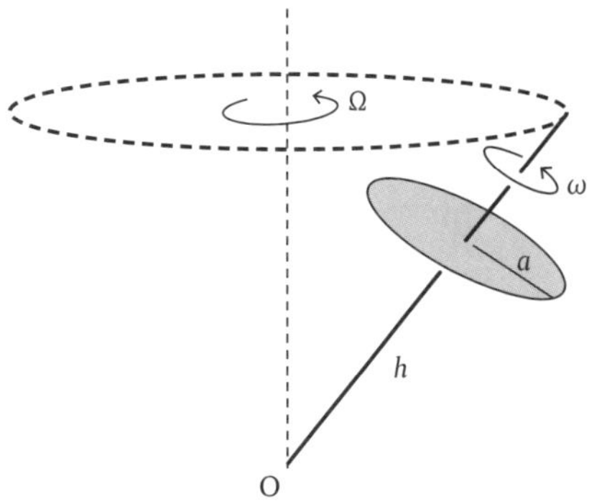

Fig. 62 The precession of a spinning top.

We may note at once, then, that the faster the top's spin ω, the slower the angular velocity of precession Ω, and this is firmly in accord with common experience. Moreover, a larger disc radius also leads to slower precession.

* * * * * * *

More generally, however, the motion of a rigid body can be far from 'simple', even when it is pivoted at a fixed point.

Some early signs of this were revealed by the great mathematician Sophie Kowalevski, who recalled in her book *A Russian Childhood* that her first experience of calculus, aged 11, came from her bedroom wallpaper, which was just tattered lecture notes on the subject that her father had acquired as a student.

Rather later, in 1888, she proved that the differential equations for a rigid body pivoted at a fixed point are fully 'integrable' only in three special cases.

(a)

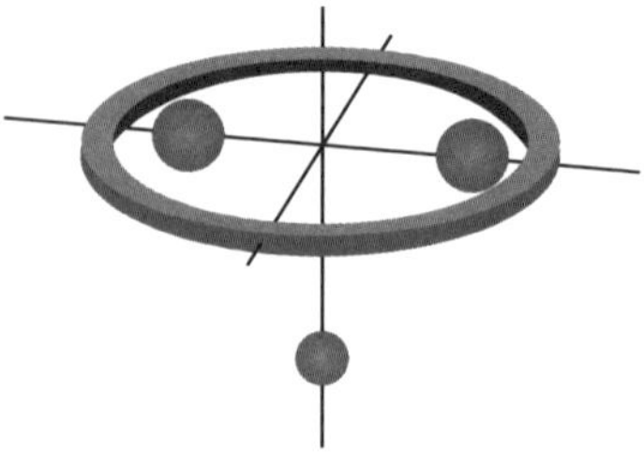

(b)

(c)

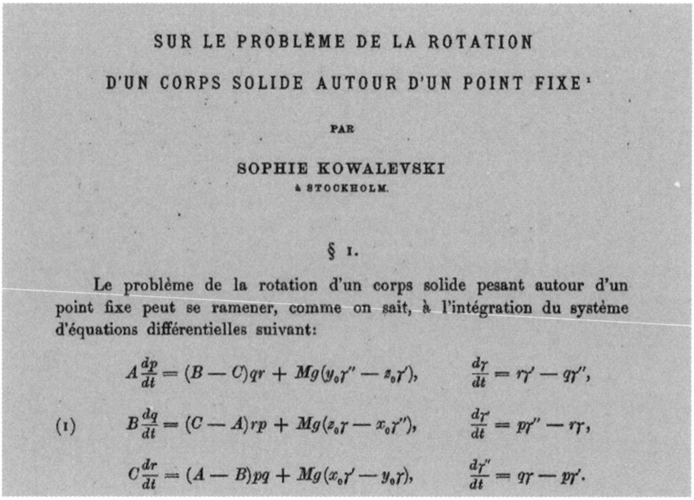

SUR LE PROBLÈME DE LA ROTATION

D'UN CORPS SOLIDE AUTOUR D'UN POINT FIXE[1]

PAR

SOPHIE KOWALEVSKI

À STOCKHOLM.

§ 1.

Le problème de la rotation d'un corps solide pesant autour d'un point fixe peut se ramener, comme on sait, à l'intégration du système d'équations différentielles suivant:

$$(1)\quad \begin{aligned} A\frac{dp}{dt} &= (B-C)qr + Mg(y_0\gamma'' - z_0\gamma'), & \frac{d\gamma}{dt} &= r\gamma' - q\gamma'', \\ B\frac{dq}{dt} &= (C-A)rp + Mg(z_0\gamma - x_0\gamma''), & \frac{d\gamma'}{dt} &= p\gamma'' - r\gamma, \\ C\frac{dr}{dt} &= (A-B)pq + Mg(x_0\gamma' - y_0\gamma), & \frac{d\gamma''}{dt} &= q\gamma - p\gamma'. \end{aligned}$$

(d)

Fig. 63 (a) Sophie Kowalevski (1850–1891). (b) An example of a Kowalevski top. (c) Signature from a copy of Poisson's *Mechanik* (1836) in the author's collection, once owned by Kowalevski herself. (d) The opening to her groundbreaking paper of 1888.

The first, studied by Euler, is when the rigid body in question is pivoted at its centre of gravity. The second, due essentially to Lagrange, is the case that we have just seen, where a top with rotational symmetry is pivoted at some point along its symmetry axis. And the third is a rather more obscure pivoted configuration which Kowalevski herself discovered (Notes, p. 239).

Loosely speaking, her work implies that, in general, there can be no 'formula' like $s = \frac{1}{2}gt^2$ to describe *for all time t* the motion of a rigid body pivoted at a fixed point. That will only ever be possible in three special cases.

And this 'nonintegrability' can in fact open the way to motions of staggering complexity, as we shall see much later in the book.

* * * * * * *

Even a rotationally symmetric top can offer major challenges if, like the tippy-top of Fig. 61, it is *not* pivoted at a fixed point, but free to move about.

A key aid to progress in that case is the fact that Euler's principle of angular momentum can be applied *about the centre of gravity* G. Nonetheless, the dynamics is still not easy, partly because friction with the floor plays a vital role in the overturning process.

For that process raises the top's centre of gravity, increasing its potential energy. This means that the top's kinetic energy must *decrease*, so that it ends up spinning more slowly. This, in turn, means that the top's angular momentum must be reduced during the overturning process, which can only be achieved by a torque about a vertical axis through its centre of gravity G. And that can only be provided by friction at the point of contact with the floor.

* * * * * * *

To end this chapter, I cannot resist pointing out the strange object in Fig. 64, which is, so far as I know, the first example of a *chiral* tippy-top.

I discovered it by accident in the late 1980s, and it gradually turns over and ends up on its stem *only if spun one way* (clockwise when viewed from above).

Fig. 64 A chiral tippy-top.

Plainly, the object, which was made for me at the time by an engineering colleague, must have some tiny lack of symmetry that is invisible to the naked eye. But explaining *this* particular phenomenon still appears to offer something of a challenge, even today (Notes, p. 240).

18

. . . AND FALLING CATS

Many years ago, during a first-year lecture on mechanics at Oxford, I invited the students to send me an explanation of how a falling cat, initially upside-down, can manage to right itself in mid-air and land on its feet.

Fig. 65 A falling cat.

I was presumably hoping for an explanation of how this could possibly be consistent with conservation of angular momentum. In the event, however, the only submission I received was the one shown in Fig. 65.

* * * * * * *

As it happens, the whole cat-turning problem has a long history. The earliest reference to it that I have found is in a book on heraldry by John Bossewell, published in 1572, where we find, in his writings on 'cattes':

> He is a cruell beaste, when he is wilde; and falleth on his owne feete from moste highe places: and noneth is hurte therewith.

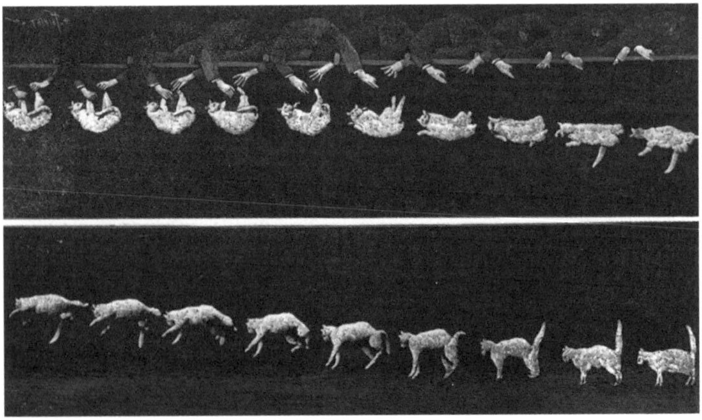

Fig. 66 From Marey's short film *Falling Cat* (1894).

A major step forward came in 1894, when Étienne-Jules Marey used an early version of high-speed photography to capture the whole phenomenon on film (Fig. 66), and this immediately prompted the French mathematician Émile Guyou to offer a theory of how the cat managed to turn itself over in mid-air.

To see something of this, however, I think we need first to look at something a bit simpler . . .

* * * * * * *

Suppose, if you will, that you are standing at rest on a sheet of ice, but that, unlike the skater in Fig. 55, you are in your carpet slippers, so that you have absolutely no prospect of generating any torque about a vertical axis through your centre of gravity. Then, whatever you do, your angular momentum about that axis will have to remain zero at all times.

Yet, surprising as it might seem, there is still a way to turn yourself round.

Start with your arms hanging down close to your body, and extend them both to your right. Then swing them round horizontally in a half-circle until they are extended to your left. While this is happening, your body will turn a little to the right, to keep your angular momentum zero.

Now drop your arms again, and move them *close to your body* from your left side to the right. This will cause your body to rotate only very slightly to the left, because the moment of inertia of your arms about a vertical axis through your centre of gravity is now much reduced. In this way, you end up having turned a little to the right.

* * * * * * *

And Émile Guyou's 1894 cat-turning theory is a bit like this. In his approach to the problem, the cat is viewed loosely as two linked rigid bodies, a fore part (including the head) and a hind part, and the motion takes place in two stages.

In the first stage, the cat draws in its fore legs, sticks out its hind legs, and rotates its fore part through as large an angle as it can. In order to conserve (zero) angular momentum, the hind part rotates in the opposite direction, but to a lesser extent, because its moment of inertia is greater.

In the second stage the cat 'untwists' by a similar manoeuvre. It extends its fore legs, draws in its hind legs, and rotates its hind part through as large an angle as it can, with its fore part counter-rotating to a lesser extent.

The end result is that its whole body has turned through a certain angle, while its angular momentum is still zero.

And the mechanism as a whole is indeed similar to that for the man standing on ice, insofar as (a) it exploits changes in moment of inertia and (b) different parts of the body counter-rotate.

* * * * * * *

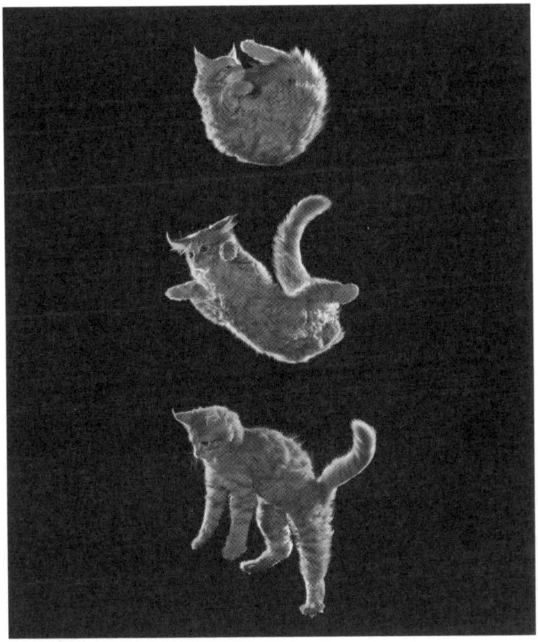

Fig. 67 A falling cat

In 1935, however, two Dutch physiologists, G. G. J. Rademaker and J. W. G. Ter Braak, proposed an entirely different mechanism for overturning that doesn't involve changes in moment of inertia at all.

This mechanism exploits instead the angle between the two portions of an articulated 'cat', as indicated in Fig. 68. These rotate about their symmetry axes in the *same* sense, without the twisting of the previous mechanism.

Now, such a motion would on its own have angular momentum about a horizontal axis through the centre of

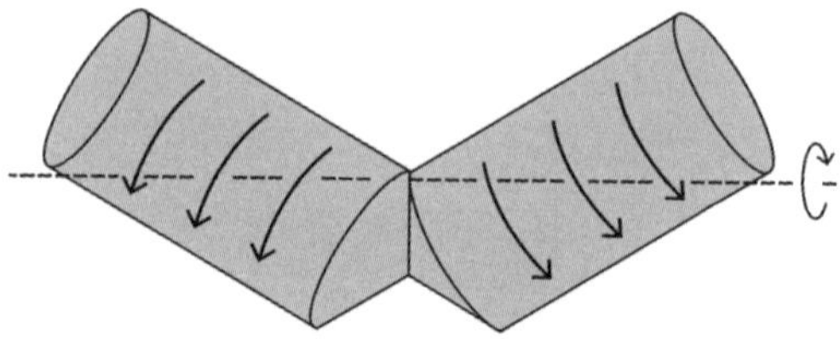

Fig. 68 An articulated 'cat'.

gravity. So, to cancel that, the *whole system counter-rotates*, as indicated by the arrow on the right of Fig. 68.

Moreover, by choosing the angle between the two portions judiciously it is possible to arrange, for instance, that as both cylinders complete one whole revolution about their own axes, the system as a whole completes half a revolution about the horizontal axis. The articulated cat that was initially upside-down is then the right way up, still with no angular momentum.

While current opinion seems to strongly favour mechanisms of this second type, I imagine that what real cats *actually* do will depend on several other things, including no doubt their general mood at the time.

19

THE NOT-SO-SIMPLE PENDULUM

One of the most iconic formulae in the whole of physics concerns the small oscillations of a simple pendulum (Fig. 69).

It was discovered, in effect, by Huygens, in his *Horologium Oscillatorium* of 1673, yet it was not until the 1730s that the new calculus methods allowed pendulum motion to be treated in the way that one can find in any modern textbook.

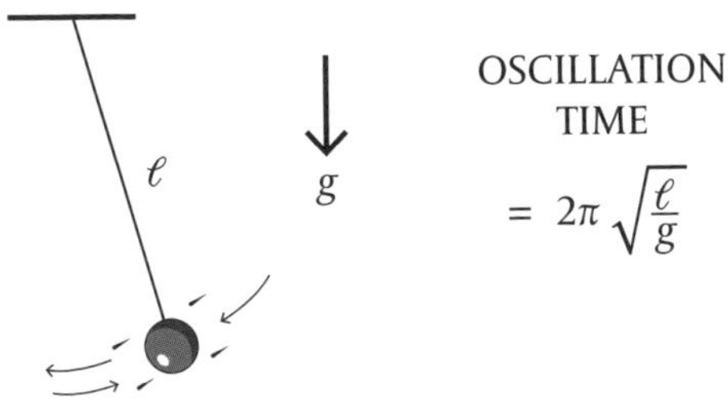

Fig. 69 The periodic time for each to-and-fro *small* oscillation of a simple pendulum.

In the case of a point mass on the end of a weightless rod of length l, the fundamental principles of mechanics then lead to the differential equation

$$\frac{d^2\theta}{dt^2} = -\frac{g}{l}\sin\theta,$$

where θ denotes the angle (in radians) with the downward vertical.

Solving this is difficult, but if the oscillations are *small* we may approximate $\sin\theta$ quite accurately by θ, and the equation then becomes

$$\frac{d^2\theta}{dt^2} = -\frac{g}{l}\theta,$$

which—most happily—is just the differential equation for the spider that we met on p. 64, but with $\omega^2 = g/l$.

It therefore follows immediately that the oscillation period must be $2\pi\sqrt{l/g}$, as claimed in Fig. 69.

'A proposition well known to Geometers'

If the oscillations are not small in size, there is one rather attractive result that appears, in fact, early in Newton's *Principia*, where he writes:

> it is a proposition well known to Geometers, that the velocity of a pendulous body in the lowest point is as the chord of the arc which it has described in its descent.

In other words, the velocity at the lowest point is proportional to the length of the straight line joining that point to the point of release (Notes, p. 240).

And Newton actually used this result in his experiments on collisions. For one of the great problems at the time was how to measure the velocities of the various objects before and after impact. To overcome this difficulty, Newton hung two pendulums side by side, so that the bobs were in contact, drew one aside, and released it so that a collision occurred.

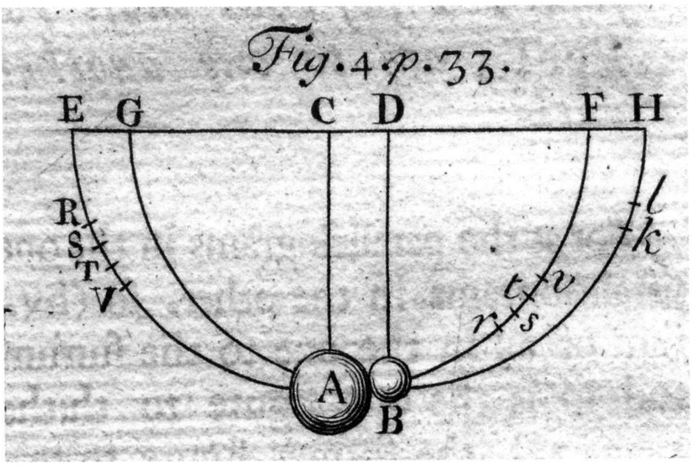

Fig. 70 Newton's use of pendulums in his experiments on collision, from *Principia*, Book 1.

And through the 'proposition' above he was able to compare two velocities essentially by simply comparing two *lengths*.

The double pendulum

A quite different way in which the 'simple pendulum' can become rather less simple is when one pendulum is suspended from another.

In 1738, for instance, Daniel Bernoulli discovered that there are then *two* 'natural' modes of oscillation (Fig. 71).

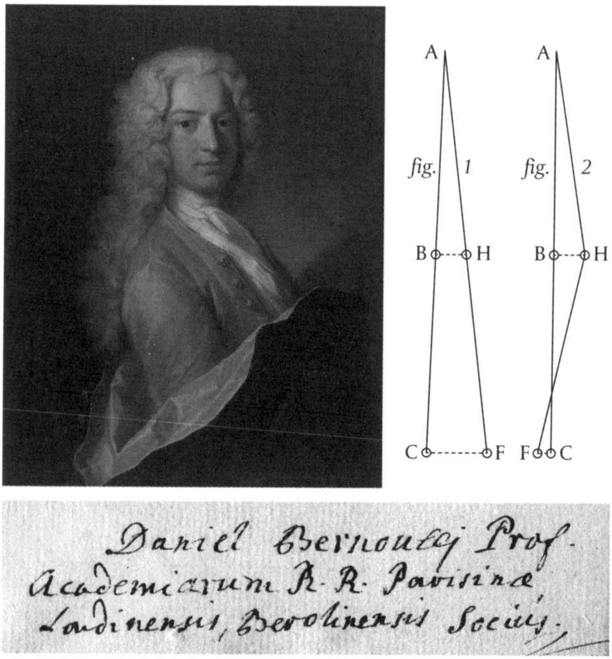

Fig. 71 Daniel Bernoulli (1700–1782), with his sketch of the two modes of oscillation of a double pendulum, and his signature on a document in the author's possession.

Suppose, for instance, that both pendulums have the same length l and bobs of equal mass. The 'slow' mode then has period

$$T = 2\pi\sqrt{\left(1 + \frac{1}{\sqrt{2}}\right)\frac{l}{g}},$$

with the two pendulums swinging in the same direction at any given moment, while the 'fast' mode has a shorter period

$$T = 2\pi\sqrt{\left(1 - \frac{1}{\sqrt{2}}\right)\frac{l}{g}}$$

and the pendulums swinging in opposite directions. In both cases, the ratio of the lower pendulum angle to the upper pendulum angle is $\sqrt{2}$.

Bernoulli confirmed all these theoretical findings with an actual experiment, using two identical lead balls connected by a steel thread. He also took the mathematical theory much further, with a hanging chain of N pendulums, finding that there are then N different natural modes.

And this whole idea of a system having many different *modes* of oscillation, each with its own characteristic frequency, gradually turned out to be an idea of enormous importance in physics more generally.

20

THE WAVE EQUATION

In virtually all the dynamics we have seen so far, the physical quantities of interest have depended on just one variable—*time*. Yet, all too often, this isn't the case.

Figure 72, for instance, shows a wave travelling down a stretched string, such as a washing-line, and it is immediately obvious that the string displacement y depends on both time t and the coordinate x.

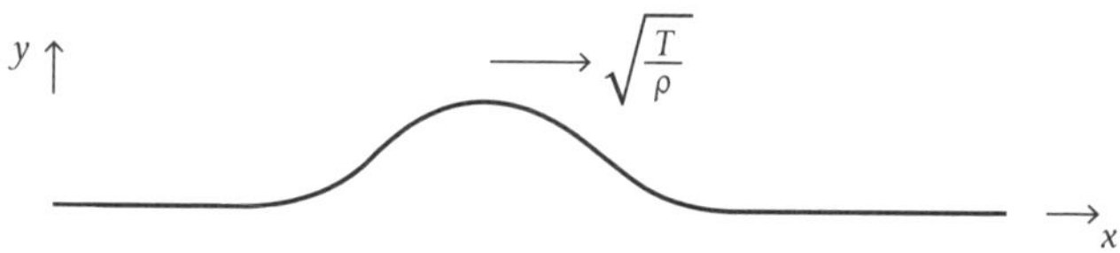

Fig. 72 A wave travelling down a stretched string.

For this reason, two different types of differentiation feature in the dynamics:

$$\frac{\partial y}{\partial t} \quad \text{and} \quad \frac{\partial y}{\partial x}.$$

The first of these is simply the rate of change of *y* with time *t* at a fixed value of *x*, and therefore denotes the sideways velocity of the string at that value of *x*.

In the same way, the second is the rate of change of *y* with *x* at a fixed time *t*, and therefore represents the instantaneous slope of the string.

And the slightly different notation '∂'—a sort of curly '*d*'—is simply to remind us that we are now engaged in so-called *partial differentiation*, where we hold some variable constant during the differentiation process itself.

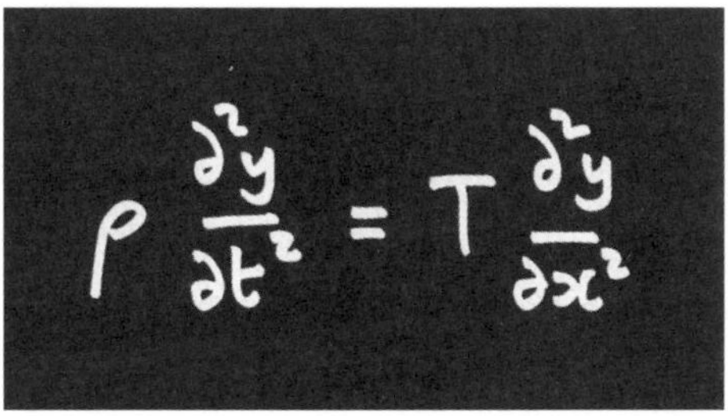

Fig. 73 The wave equation for a stretched string.

In 1747 Jean Le Rond d'Alembert showed that small-amplitude waves on a stretched string are governed by the *wave equation* in Fig. 73, where *T* denotes the tension in the string and ρ its mass per unit length.

Here, the right-hand side comes from the net force on a small element of the string, and the left-hand side from the product of its mass and acceleration.

And the major conclusion of d'Alembert's 1747 paper was that disturbances can travel along the string, in either direction, without change of shape, at a

$$\textit{wave speed} = \sqrt{T/\rho}.$$

In particular, then, and as one might expect, the greater the tension in the string, the faster the waves travel.

And while d'Alembert could not possibly have known it, this very same equation (with different physical constants) was to keep reappearing throughout the history of dynamics. This happened first in connection with surface waves on water, then with sound waves, but most famously perhaps in connection with electromagnetic waves, where it played a key part in James Clerk Maxwell's 1865 discovery that light is an electromagnetic phenomenon (Fig. 74).

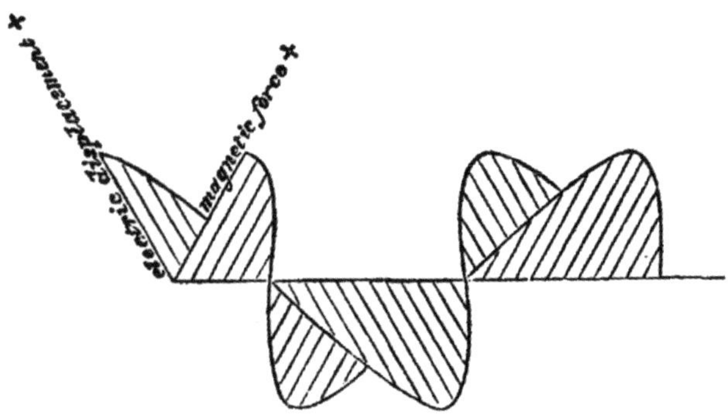

Fig. 74 An electromagnetic wave, from Maxwell's *Treatise in Electricity and Magnetism* (1873).

* * * * * * *

One key property of the wave equation in Fig. 73 is that it is *linear*, which means, above all, that the sum of any two solutions will also be a solution of the equation.

And one immediate consequence is a neat trick, due to Euler, for deducing how a wave gets *reflected* when one end of the string is fixed.

The idea is to consider a different problem, where the string is *not* fixed at the point in question, but extends freely instead beyond that point (Fig. 75). And the real masterstroke is to then imagine *another* wave, equal and opposite, travelling in along the other half of the string.

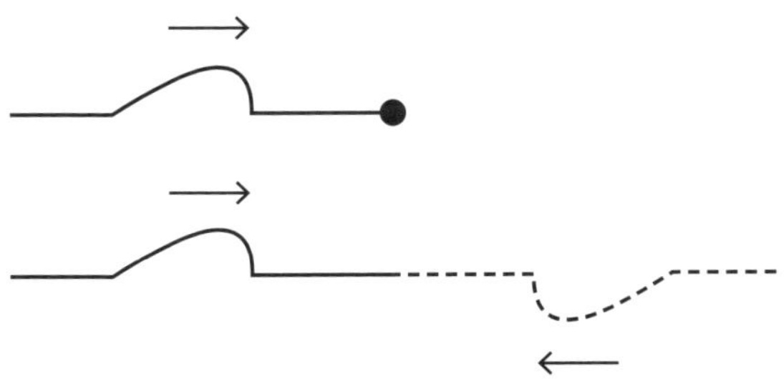

Fig. 75 A reflection problem, by Euler's 'method of images'.

For the reasons just mentioned, the total displacement of the string will then be simply the sum of the two individual waves. But these will plainly always cancel each other out at the point where the fixed end used to be, giving a permanently zero displacement there, just as if that particular point of the string *were* fixed.

The motion to the left of that point will therefore give us the solution to the original problem, so explaining why a travelling wave gets reflected from a fixed end *upside-down*.

And a casual experiment with a washing-line confirms that this is, indeed, exactly what happens.

* * * * * * *

Not surprisingly, when *both* ends are fixed, as on a violin or guitar, solutions of the wave equation have major applications to music.

They reveal, in particular, a whole series of natural 'modes' of vibration (Fig. 76), each with its own distinctive

$$\textit{vibration frequency} = \frac{N}{2l}\sqrt{\frac{T}{\rho}},$$

where l denotes the length of the string and N is a positive whole number.

The first of these 'modes', with $N = 1$, is known as the fundamental, and such that the whole string vibrates up and down, with all parts moving in the same direction at any given moment. And, notably, the frequency is inversely proportional to l, the length of the string that is actually vibrating, which explains why pressing down on a higher fret on a guitar produces a higher note.

In the $N = 2$ mode, on the other hand, the two halves of the string move in opposite directions at any given

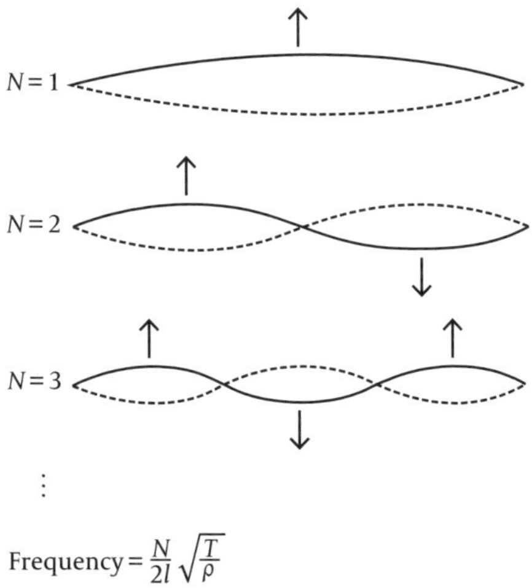

Fig. 76 Natural modes of vibration of a stretched string with fixed ends.

moment, and vibrate at twice the fundamental frequency, producing a note one octave higher.

And this whole idea of different modes of oscillation, at discrete frequencies, was also to re-emerge many times in the history of dynamics, most famously perhaps in connection with Erwin Schrödinger's partial differential equation for quantum mechanics in 1926.

For in Niels Bohr's theory of the hydrogen atom the energy levels of the electron were *quantized*, and proportional to $1/N^2$, where N is a whole number. And Schrödinger was able to show that this quantization could be explained much as d'Alembert's wave equation can account for the various different modes in Fig. 76.

Fig. 77 The $N = 4$ mode of vibration, from *The World of Sound*, by W. H. Bragg (1920).

And just in case you wonder whether I am stretching a point here, here are a few words from Schrödinger himself, writing in 1926:

> I wish to consider . . . the hydrogen atom, and show that . . . when integralness does appear, it arises in the same natural way as it does in the case of the node-numbers of a vibrating string.

21

'THE LAWES . . . OF SPRINGINESSE'

At this point in the story, we need to turn the clock back for a moment to November 1679 and Robert Hooke's letter to Isaac Newton. For in it Hooke also invited any comments Newton might have on his

> hypothesis of the lawes or causes of Springinesse.

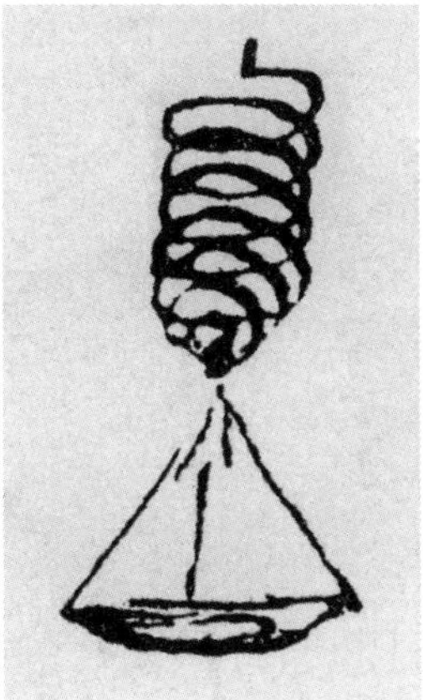

Fig. 78 A sketch of Hooke's 'philosophicall spring scales', from his diary, 21 August 1678.

This was of course what we now know as Hooke's law, namely that the force exerted by a spring is proportional to its extension.

Hooke had, in fact, been working on the topic for quite some time. A diary entry for 6 October 1675, for instance, reads:

> Walked into the Park with Sir Chr. Wren. The King calld me to him, bid me show him experiment. Followed him through tennis court garden &c, into closet. Shewd him the Experiment of Springs. He was very well pleased.

* * * * * * *

Yet, it was soon to be problems of elastic *bending* that would really propel the subject forward, especially when James Bernoulli proposed the problem of the *elastica* in 1691. This is concerned with determining the shape taken up by a thin elastic strut when conditions are imposed on its ends.

The deepest and most extensive investigation of this problem was eventually undertaken by Euler, who carefully classified the various possibilities in 1744, some of which can be seen in Fig. 79.

* * * * * * *

While static problems of elasticity could be challenging enough, investigations into the *dynamics* of elastic bodies also began quite early, and in December 1734 Daniel Bernoulli wrote to Euler about the small transverse

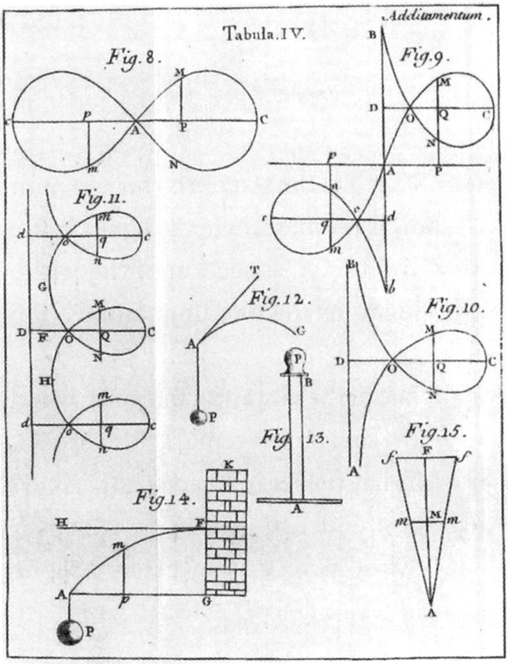

Fig. 79 Euler's researches on the elastica, from the appendix to his *Methodus inveniendi lineas curvas maximi minimive proprietate gaudentes* of 1744.

vibrations of an elastic strut with one end fixed in a wall, adding

> I am not very pleased with my solution.

Euler had, in fact, already made some progress on the problem, and showed the following year that the vibrations take place with

$$\textit{frequency} \propto \frac{1}{l^2}\sqrt{\frac{B}{\rho}}.$$

Here l is the length of the strut, ρ its mass per unit length, and B its 'bending stiffness', which depends only on the material used and the cross-section of the strut. And, not surprisingly, the larger the bending stiffness B, the faster the vibration.

The dependence of the vibration frequency on the length of the strut l is rather more remarkable, especially if we compare it with that for a stretched string. For, as we have seen already, the frequency in that case is inversely proportional to l, while here it is inversely proportional to *l squared.*

Fig. 80 Vibrations of a square plate caused by a violin bow.

And when, some decades later, the German physicist and musician Ernst Chladni performed his famous experiments on the vibration of elastic plates (Fig. 80), with sprinkled sand gathering neatly over the nodal lines, it was found once again that the frequency of any particular mode is inversely proportional to the square of the horizontal length scale.

But while Chladni's experiments rather caught the public imagination at the time, the subject of elasticity was, in fact, about to take a whole new turn, as we will see later in the book.

22

THE BEST OF ALL POSSIBLE WORLDS?

By the middle of the eighteenth century some aspects of mechanics were heading in a very different direction.

Much earlier, in 1662, Pierre de Fermat had proposed that when light is refracted at a plane boundary from one given point to another, it always takes the path of least time.

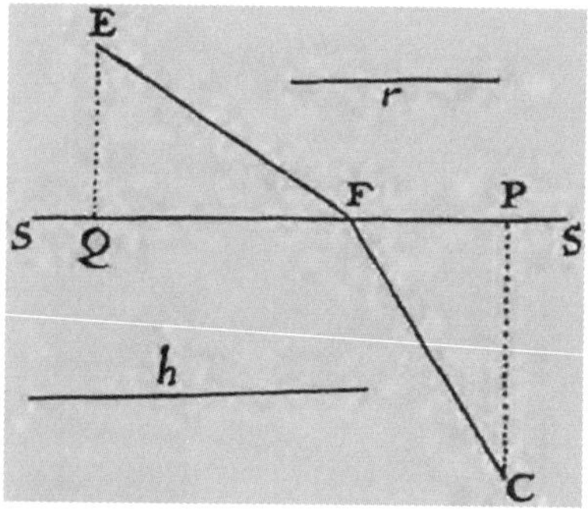

Fig. 81 The refraction of light, from Leibniz's groundbreaking paper on calculus (1684).

He had even made the sweeping—if somewhat vague—pronouncement that

> Nature operates by the simplest and most expeditious ways and means,

and in due course this fed into Leibniz's famous (but challenging) philosophical proposition of 1710 that we might all be living in 'the best of all possible worlds'.

It became only natural to enquire, then, whether some such maximum or minimum principle might be operating in *mechanics*, and in 1744 Pierre de Maupertuis proposed the following:

> When a change occurs in Nature, the Quantity of Action necessary for that change is as small as possible.

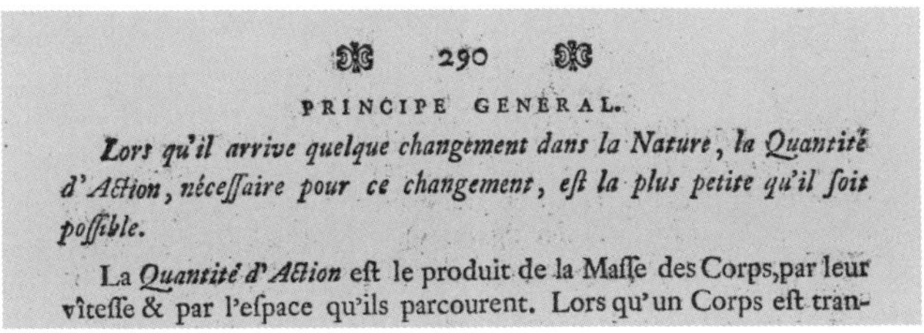

290

PRINCIPE GENERAL.

Lors qu'il arrive quelque changement dans la Nature, la Quantité d'Action, néceſſaire pour ce changement, eſt la plus petite qu'il ſoit poſſible.

La *Quantité d'Action* eſt le produit de la Maſſe des Corps, par leur vîteſſe & par l'eſpace qu'ils parcourent. Lors qu'un Corps eſt tran-

Fig. 82 Maupertuis's principle of least action, as published in his *Les Loix du mouvement et du repos déduites d'un principe metaphysique* (1746).

Moreover, Maupertuis took the view that if nature did indeed operate in this way, it would provide

> proof of the existence of Him who governs the world.

But there were problems.

* * * * * * *

The first difficulty was that Maupertuis' own inflated pronouncements about his new principle immediately plunged him into a very unpleasant priority dispute with supporters of Leibniz. Even the novelist Voltaire got involved, attacking Maupertuis's philosophy mercilessly in his *Diatribe du Docteur Akakia* of 1752, with barbs like

> We ask forgiveness of God for having pretended that there is only proof of his existence in A + B divided by Z, etc.

(a) (b)

Fig. 83 (a) Pierre de Maupertuis (1698–1759). (b) Voltaire (1694–1778).

Another problem was that Maupertuis defined his 'Quantity of Action' in general only rather loosely. It has dimensions of energy × time, but a clear and general definition only really emerged after further work, first by Euler and then, much later, by William Rowan Hamilton in Ireland.

In the end, however, the biggest difficulty was that the 'least action' idea, as originally conceived, simply isn't always true. As Hamilton himself wrote in 1833:

> the quantity pretended to be economised is in fact often lavishly expended.

Yet the whole episode gradually paved the way for an approach to many aspects of mechanics which was entirely new in two very different ways.

First, in this new way of looking at things, the motion of a system over time is often conceived as taking place in order to achieve some overall *purpose*, rather than being the result of moment-to-moment cause and effect. And this called for a wholly new branch of mathematics, largely invented by Euler and Joseph-Louis Lagrange, called the calculus of variations.

But the second novel feature was the way that ideas of *energy*, rather than force or momentum, took centre stage . . .

* * * * * * *

A high point with the new approach came with the publication of Lagrange's celebrated *Mécanique Analytique* (1788), in which the central concept is actually the *difference* between the kinetic energy T of a system and its potential energy V:

$$L = T - V.$$

(a) (b)

Fig. 84 (a) Lagrange's equations of mechanics. Despite appearances, they are *n ordinary* differential equations governing how the 'generalized coordinates' $q_1 \ldots q_n$ depend on time t (see Notes, p. 243). (b) Joseph-Louis Lagrange (1736–1813).

One advantage of the Lagrange approach is that it offers considerable freedom in the choice of coordinates to represent the configuration of the system at any particular time. But there is another, rather deeper, advantage, too.

A major feature of modern theoretical physics is the way in which conservation laws of various kinds emerge from so-called *symmetries* of a system, and this is largely through the work of the great German mathematician Emmy Noether (1882–1935). For in a deep theorem of 1918 she revealed an intimate link between the two. And the Lagrange equations of Fig. 84 provide an early example.

Suppose, for instance, that the 'symmetry' arises in this case because L is independent of one of the 'generalized coordinates', q_1, say. Then $\frac{\partial L}{\partial q_1} = 0$, and it follows immediately from the Lagrange equations themselves that $\frac{\partial L}{\partial \dot{q}_1} = constant$, so we have a conserved quantity, independent of time.

More significantly still, invariance of L with respect to translation turns out to lead to conservation of momentum, while invariance with respect to rotation leads to conservation of angular momentum.

And all such conservation laws are of great value in practice, for dynamics is all about making sense of change, and it is often rather easier to do this if you know a few things that *don't* change as time goes on.

23

INSTABILITY

I was once at a theatre in Newcastle, waiting to go on stage as part of a show called *Maths Inspiration*, when one of the other performers whispered to me in the wings that, ten years earlier, he had been a first-year student in the front row of my dynamics lectures at Oxford.

And he said that he remembered, in particular, one experiment I had done to demonstrate *instability*, which involved hurling a lightweight cardboard box high into the air.

Somewhat intrigued by this, I asked him why he thought he remembered that particular experiment.

'That's easy,' he said, 'When the box came down again, it hit me on the head.'

* * * * * * *

To see the instability in question, all you need is indeed a lightweight cardboard box, provided that its three dimensions are all rather different, as in Fig. 85.

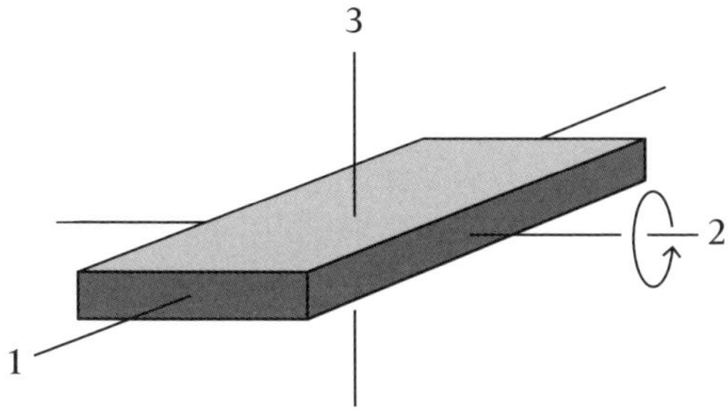

Fig. 85 A spinning box instability.

For if you throw the box up into the air so that it spins about its axis of minimum moment of inertia (1) or maximum moment of inertia (3), the motion is quite stable. Yet, if you try to set it spinning about the 'intermediate' axis (2), you will find that the original rotation is thoroughly unstable and that a spectacular tumbling motion quickly develops instead.

* * * * * * *

Dynamics is full of examples like this, though the stability of any particular state typically depends on various parameters of the system, often because some destabilizing mechanism is opposed by a stabilizing one.

One of the earliest examples, due to Euler in 1744, is the criterion for a straight elastic strut to buckle under a weight (Fig. 86):

$$F > \pi^2 B/L^2.$$

Here F denotes the downward force due to the weight, L the length of the strut, and B its 'bending stiffness', which, as I mentioned in Chapter 21, depends only on the material of the strut and the geometry of its cross-section. And when F exceeds the critical value, the slightest disturbance to the system will cause the strut to buckle either to the left or to the right, depending on the nature of the disturbance.

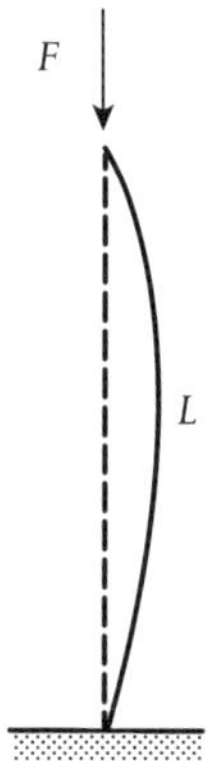

Fig. 86 Euler buckling of an elastic strut.

Most notably, perhaps, the critical value of F is inversely proportional to L^2, so that as Euler himself observes:

> A column twice as high will thus be able to bear but the fourth part of the weight.

* * * * * * *

Another good example is the spinning top of Fig. 87, where gravity acts to destabilize the fully upright state while rotation acts to stabilize it.

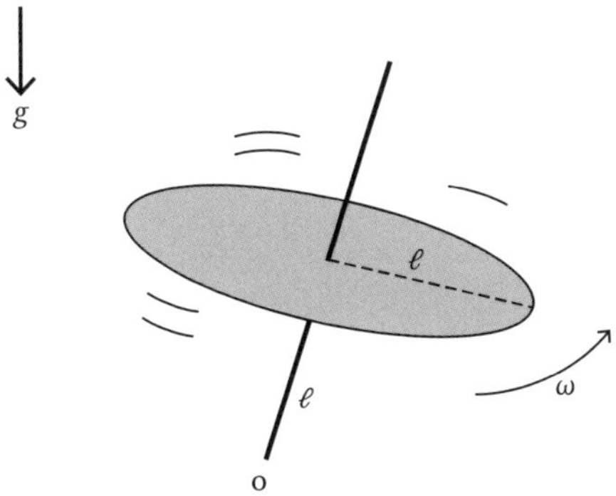

Fig. 87 A spinning top.

And it turns out in this case that the rotation 'wins' if

$$\omega > 2\sqrt{\frac{5g}{l}},$$

where ω denotes the top's angular velocity of spin about its symmetry axis.

* * * * * * *

A rather more mysterious example of instability is provided by the toy in Fig. 88, called a *rattleback*.

It consists of a wooden boat, constructed with all the symmetry you might expect, and two turtles attached to its top. And if we place the object on a flat surface and spin it so that the turtles are going forward, the object spins happily until it is slowed by friction.

Fig. 88 A rattleback.

Yet if we spin it so that the turtles are going *backward*, the motion soon becomes violently unstable, and the rotation gives way to a 'rattling' motion in which the turtles just rock up and down without spinning.

And, as if this were not enough, the rattling itself then gives way to a slow spinning motion *in the opposite direction to the original!*

While the whole phenomenon has now been known for at least a hundred years, learned papers still appear in the scientific literature on the precise mechanisms by which all this happens.

Yet even deeper mysteries of dynamics emerge as soon as we turn to the motion of a *fluid* . . .

24

FLUID DYNAMICS

The first major application of dynamical ideas to *fluid* motion came in 1738 with the publication of Daniel Bernoulli's *Hydrodynamica*.

He was interested, in particular, in the relation between the pressure p associated with the fluid motion and the flow speed v. And today it is only natural to look first for the famous theorem that now bears his name:

Bernoulli's theorem

In the steady flow of a nonviscous, incompressible fluid of constant density ρ,

$$p + \frac{1}{2}\rho v^2 = constant$$

along a streamline.

In short, then, along any particular streamline, the pressure p associated with the fluid motion is large where the flow speed v is small, and vice versa. And in case this seems a little strange at first, it does, in fact, make sense.

Imagine, for instance, water flowing in a narrowing channel, as in Fig. 90. As water is essentially incompressible, it

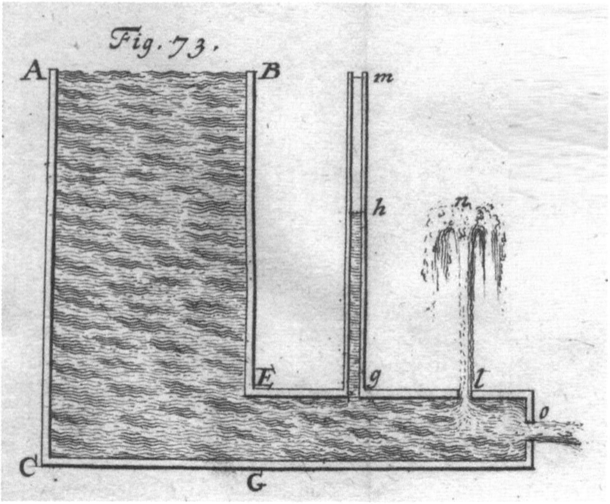

Fig. 89 From Daniel Bernoulli's *Hydrodynamica* (1738).

has to flow faster at B than at A, because the channel is narrower there. According to Bernoulli's theorem, then, the pressure at A should be higher than the pressure at B.

And this does indeed make sense, for it means that the pressure on the back of a small fluid element is slightly higher than the pressure on its front, which is precisely what causes the *acceleration* of that particular element as it moves along a streamline from A to B.

Daniel Bernoulli's theorem, then, *is* in keeping with one of the great principles of dynamics, yet if we look for it in the *Hydrodynamica*, it is almost unrecognizable (Notes, p. 244).

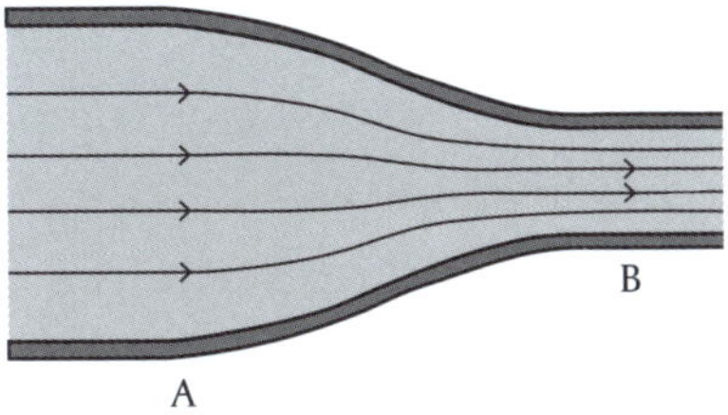

Fig. 90 Flow in a narrowing channel.

And, in truth, the theorem *as we know it today* owes just as much to his father Johann and to Euler.

* * * * * * *

It was not until later, in a paper of 1755, that Euler obtained the first general differential equations of fluid motion.

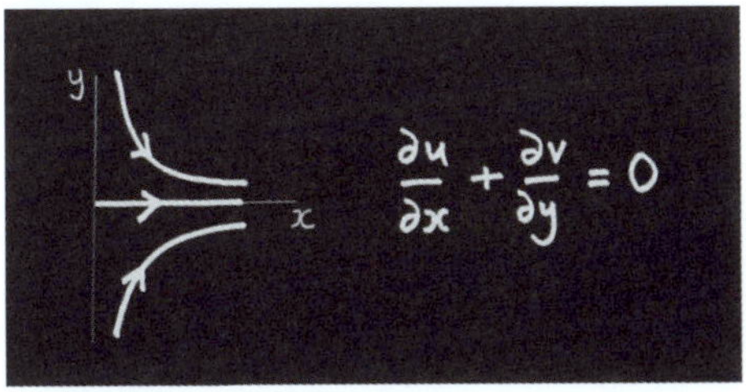

Fig. 91 The incompressibility equation, for a flow with two velocity components (u, v). The symbol $\frac{\partial u}{\partial x}$ means, loosely, a small increase in u divided by a corresponding small increase in x *while keeping* y *constant*, and $\frac{\partial v}{\partial y}$ can be interpreted similarly.

To get something of the flavour of the Euler equations, consider first how the incompressibility of a fluid like water can be expressed in mathematical terms.

Suppose, for instance, that the flow is two-dimensional, so that there are just two velocity components (u, v) depending on coordinates x, y.

Incompressibility then turns out to be captured very neatly—and symmetrically—by the partial differential equation in Fig. 91.

This then has to be supplemented by three further equations which come from applying the principle of linear momentum—or *force = mass × acceleration*—to a small element of fluid.

It is important to note, however, that Euler's equations are for a *nonviscous* fluid. This means that the internal forces across the boundary between two adjacent elements of fluid are due to the pressure alone; no tangential stresses are involved.

* * * * * * *

Euler's equations of nonviscous fluid mechanics represent one of the great scientific highlights of the eighteenth century.

In other respects, however, the subject got off to a rather shaky start. For while Euler derived what we now know as 'Bernoulli's theorem' from his general equations of motion, he was unable to deduce from those same equations much else of a genuinely practical nature.

Worse still, just a few years earlier, d'Alembert had studied the problem of uniform, nonviscous flow past a sphere.

And he had found that the corresponding flow was symmetric fore and aft of the body, with the streamlines parting smoothly around the front and then closing up again, just as smoothly, at the back (Fig. 92a).

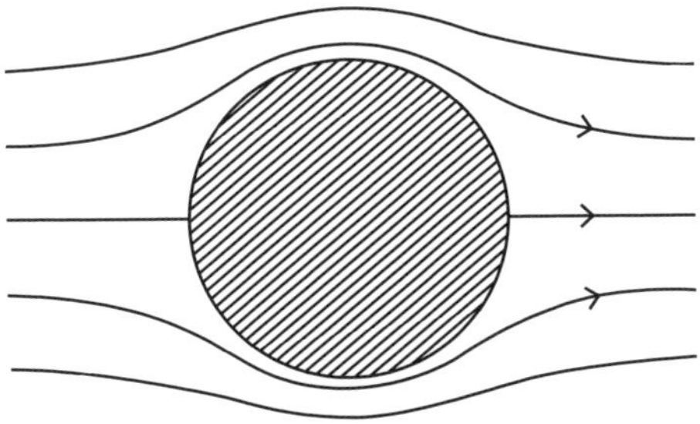

(a)

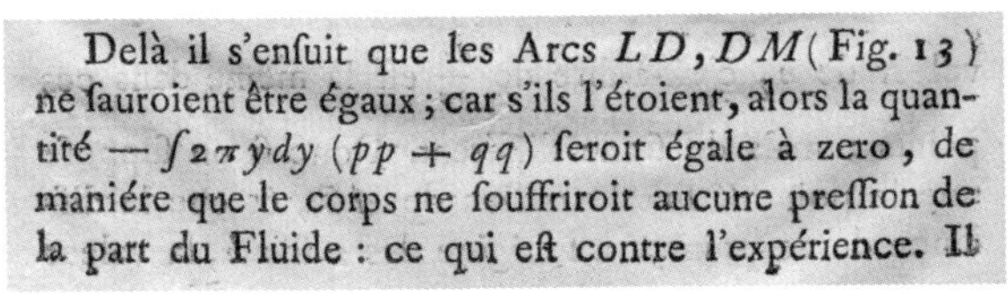

Delà il s'enſuit que les Arcs *LD*, *DM* (Fig. 13) ne ſauroient être égaux ; car s'ils l'étoient, alors la quantité $-\int 2\pi y dy\ (pp + qq)$ ſeroit égale à zero, de maniére que le corps ne ſouffriroit aucune preſſion de la part du Fluide : ce qui eſt contre l'expérience. Il

(b)

Fig. 92 (a) Nonviscous flow past a sphere. (b) D'Alembert's paradox, as it appears in his *Essai d'une Nouvelle Théorie de la Résistance des Fluides* of 1752.

As the flow itself was symmetric, Bernoulli's theorem then implied that the pressures on the back of the object were the same as those on the front, so that the net force, or *drag*, on the object was *zero*.

D'Alembert acknowledged at once that this was 'contrary to experience', writing

> I do not see, I admit, how one can satisfactorily explain by theory the resistance of fluids. On the contrary, it seems to me that the theory, in all rigour, gives in many cases zero resistance; a singular paradox which I leave to future Geometers for elucidation.

And, in the event, it would take over 150 years for d'Alembert's 'paradox' to be fully explained, as we will see later in the book.

25
VISCOUS FLOW

The mathematical study of viscous flow began in Newton's *Principia*, though not in the way that many people seem to think.

For viscosity is traditionally introduced in textbooks through flow down a straight channel, yet Newton's fundamental hypothesis on the subject is actually contained wholly within a section on *circular* fluid motion, essentially because his main aim is to demolish René Descartes' vortex theory of the heavens (Fig. 93).

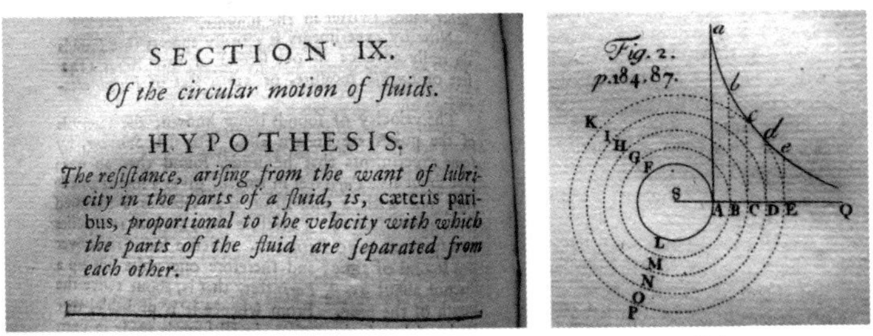
SECTION IX.
Of the circular motion of fluids.

HYPOTHESIS.

The resistance, arising from the want of lubricity in the parts of a fluid, is, cæteris paribus, *proportional to the velocity with which the parts of the fluid are separated from each other.*

(a) (b)

Fig. 93 (a) Newton's hypothesis about viscous forces. (b) His treatment of the circular motion of a viscous fluid.

Moreover, Newton's very first deduction from that hypothesis—about the viscous flow generated outside a rotating cylinder (Fig. 93b)—is actually wrong, providing just one small indication of how rudimentary fluid mechanics was at the time, even in the hands of a great master (Notes, p. 246).

* * * * * * *

In fairness, there can be little doubt that the standard textbook way of introducing the idea of viscosity, with straight streamlines, captures essentially what Newton had in mind (Fig. 94).

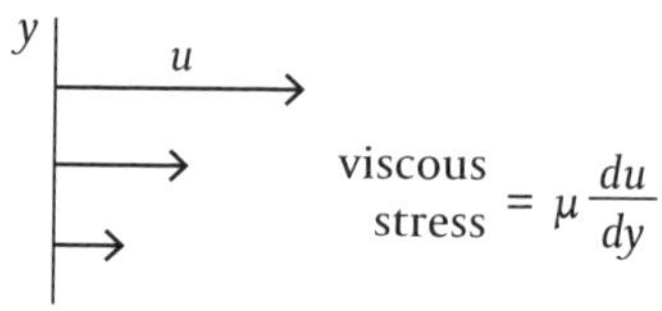

Fig. 94 Viscous stress.

Each layer of fluid moving with velocity u then exerts a tangential stress (i.e. force per unit area of contact) on the layer immediately below, and this stress is proportional to the velocity gradient. So, the more a fluid is *sheared*, the greater the value of du/dy, and the greater the viscous forces. And the proportionality constant μ is called the coefficient of viscosity.

While this might all seem simple enough, it was not until 1845 that the general differential equations of viscous fluid motion emerged, from fundamental work by Claude-Louis Navier in France and George Stokes in Cambridge.

The Navier–Stokes equations

For an incompressible fluid of constant density ρ these amount to four partial differential equations for four 'unknowns', namely the three components of velocity u, v, w and the pressure p.

One of them, expressing incompressibility, is just a natural generalization of the one that we met in Chapter 24:

$$\frac{\partial u}{\partial x} + \frac{\partial v}{\partial y} + \frac{\partial w}{\partial z} = 0.$$

Fig. 95 Sir George Stokes (1819–1903).

The other three come from the three components of *force* = *mass* × *acceleration* applied to a small fluid element. These all have exactly the same structure, and the first, representing the x-component, is

$$\frac{\partial u}{\partial t} + u\frac{\partial u}{\partial x} + v\frac{\partial u}{\partial y} + w\frac{\partial u}{\partial z}$$
$$= -\frac{1}{\rho}\frac{\partial p}{\partial x} + \frac{\mu}{\rho}\left(\frac{\partial^2 u}{\partial x^2} + \frac{\partial^2 u}{\partial y^2} + \frac{\partial^2 u}{\partial z^2}\right).$$

The left-hand side represents the x-component of acceleration of the element, while the right-hand side denotes the corresponding component of force (per unit mass) due to the pressure gradient and viscous stresses (Notes, p. 246).

If $\mu = 0$, the viscous terms disappear, and the equations then revert to those introduced by Euler in 1755.

The no-slip condition

It is not only the differential equations which change when viscosity is included in the mathematical model.

In the nonviscous theory, the fluid must be allowed to *slip* freely along solid boundaries, but if the viscosity μ is not zero, there is no such slip, and the tangential flow speed must then match that of the boundary itself.

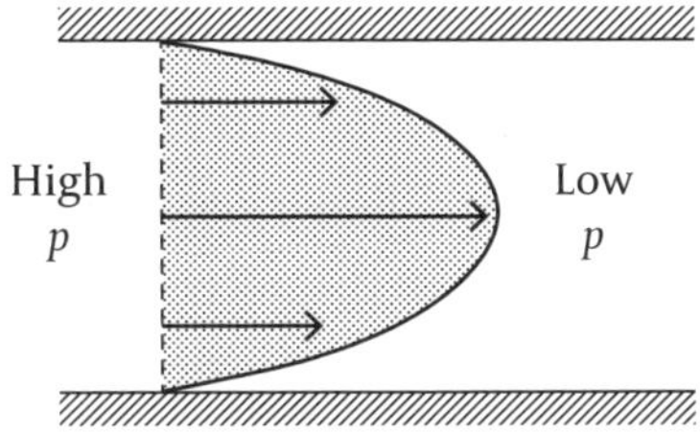

Fig. 96 Viscous flow in a pipe of circular cross-section.

This is well illustrated by a simple exact solution of the Navier–Stokes equations for steady flow down a pipe, for if we denote the pipe's radius by a, the velocity is then given by

$$u = u_{max}\left(1 - \frac{r^2}{a^2}\right),$$

where r denotes distance from the centre line.

In particular, then, the flow is fastest on the centre line itself, and the velocity is zero on the pipe wall $r = a$, in accord with the no-slip condition (Fig. 96).

The Reynolds number

More generally, the Navier–Stokes equations pose great mathematical challenges, even today. This is not simply because the equations themselves contain a lot of terms, but because so many of those terms, like $u\frac{\partial u}{\partial x}$, are *nonlinear*, involving *products* of physical quantities that we are trying to find.

One thing we can do quite easily, however, is make some rough order-of-magnitude estimates of the sizes of the various terms. For if U denotes a typical flow speed and L a typical length scale of the motion, then the acceleration terms in the Navier–Stokes equations will be of order U^2/L, while the viscous terms turn out to be typically of order $\mu U/\rho L^2$.

The ratio of these two order-of-magnitude estimates is

$$R = \frac{\rho UL}{\mu},$$

which is known as the *Reynolds number*.

And it is therefore natural to expect that flows at low and high Reynolds numbers will have very different characteristics.

Some typical values of the Reynolds number, R

Flow past wing of jumbo jet	5×10^7
Stirred cup of tea	5000
Swimming spermatozoon	0.01
Syrup draining from a spoon	0.002

A classic experiment

As it happens, the Reynolds number itself originally entered fluid mechanics in a rather different way, through a classic experiment of 1883 on the stability of flow down a circular pipe (Fig. 97).

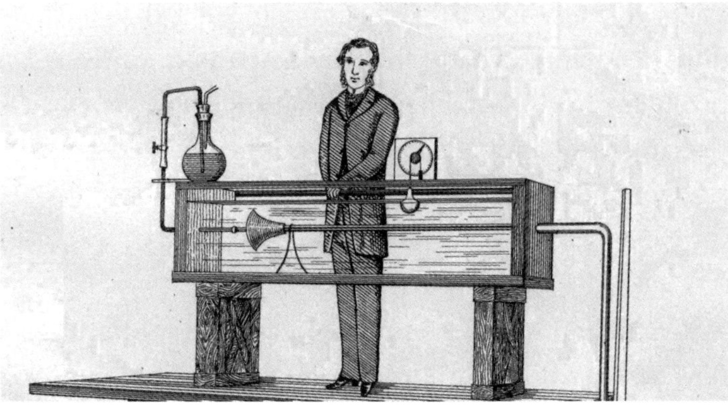

Fig. 97 Detail from Osborne Reynolds' own drawing of his famous experiment on flow down a pipe (1883).

In this particular context, the Reynolds number is typically defined more precisely as $R = \rho Ud/\mu$, where U is the mean velocity down the pipe and d the pipe diameter.

And a widespread notion, which I remember being taught in my schooldays, is that the flow changes from the smooth 'laminar' form of Fig. 96 to *turbulence* when R is gradually increased beyond about 2000.

Yet it turns out that this is only actually true if the experiment is done rather carelessly. For, in his 1883 paper, Reynolds himself reported that

> the critical velocity was very sensitive to disturbance in the water before entering the tube

and by taking great care to minimize such disturbances he was able to keep the flow stable and laminar up to values of R approaching 13,000. Modern experiments have achieved much more dramatic results of a similar kind.

The real significance of the figure of 2000, then, is quite different, namely that if the flow down a pipe already *has* turbulence, it is only when you *reduce R* below about 2000 that the turbulence disappears, and the flow reverts to its smooth or laminar form.

* * * * * * *

Notwithstanding all these developments, uniform fluid flow *past an object* continued to be a puzzle throughout the nineteenth century, and d'Alembert's paradox remained unresolved, with nonviscous theory continuing to give completely wrong answers, even when the viscosity of the fluid seemed so small as to be negligible.

And what made it all the more puzzling was that the same nonviscous theory worked very well indeed in connection with many other aspects of fluid mechanics, such as wave motion . . .

26

'I GOT INSTANTLY ON HORSEBACK'

One of the most important ideas of modern physics arose from a chance observation by the Victorian engineer John Scott Russell on the Edinburgh–Glasgow canal in August 1834.

He saw a fast-moving vessel come to a sudden stop and generate a large, single wave which then travelled a long

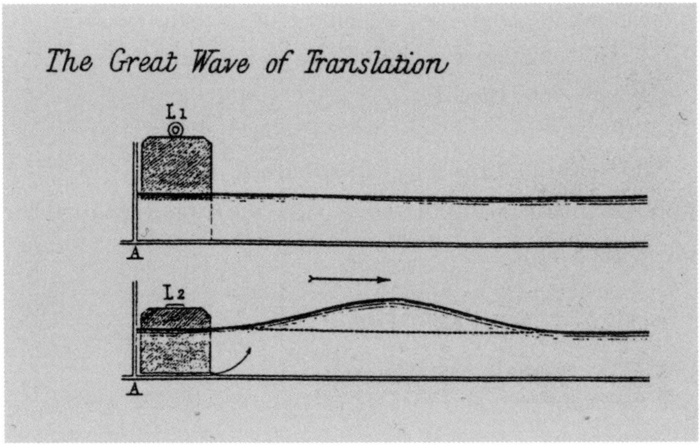

Fig. 98 John Scott Russell's sketch of an experiment with solitary waves.

way down the canal without change of form. As he recalled later:

> I immediately . . . attempted to follow this wave on foot, but finding its motion too rapid, I got instantly on horseback and overtook it in a few minutes, when I found it pursuing its solitary path with a uniform velocity along the surface of the fluid. After having followed it for more than a mile, I found it subside gradually, until at length it was lost among the windings of the channel.

The original wave was apparently some 30 feet in length and rather more than a foot in height. Russell then conducted extensive laboratory experiments on such 'solitary waves' (Fig. 98). And by 1837, when he reported the phenomenon to the Royal Society of Edinburgh, he observed that

> A large or high wave had a greater velocity than a small one. When a small wave preceded a large one, the latter invariably overtook the former.

And to understand why Russell found all this so puzzling, and worthy of careful investigation, we must first turn the clock back a bit . . .

Small-amplitude water waves

The earliest studies of water waves based on Euler's equations of fluid mechanics assumed that the amplitude, or height of the waves, was very small. This meant that nonlinear terms in the differential equations could be neglected, which led to major simplifications in the theory.

The first study of this kind was by Lagrange in 1781, who considered waves in shallow water of depth h. And he showed that the wave speed is then

$$c = \sqrt{gh}.$$

Notably, this speed is determined purely by the (undisturbed) depth of fluid h, and any such wave travels without change of shape, in a manner reminiscent of the waves on a stretched string in Chapter 20.

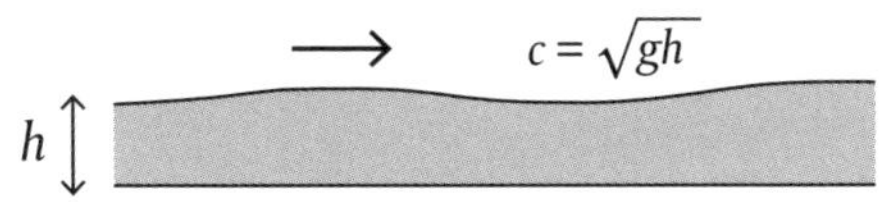

Fig. 99 Small-amplitude waves in shallow water.

* * * * * * *

Lagrange's theory assumes that the wavelength λ—i.e. the distance between successive crests—is much greater than the depth h. Somewhat later, in 1816, Siméon Denis Poisson examined the opposite extreme, when λ is much *smaller* than h, and found that the wave speed is then given instead by

$$c = \sqrt{g\lambda/2\pi},$$

so that the wave speed is proportional to the square root of the wavelength. And, as it happens, this had, in fact, been anticipated by Newton in his *Principia* of 1687, on the basis of a somewhat loose analogy with pendulum motion (Fig. 100).

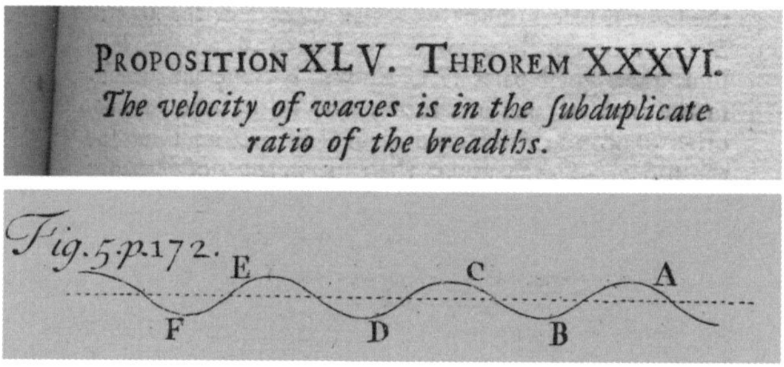

Fig. 100 Newton's treatment of surface waves on deep water, from his *Principia*, Book 2.

The fact that long waves travel faster changes everything, and means that surface waves on deep water are very different to waves on a stretched string, because they now change in form as they travel, by spreading out in a process called *dispersion*.

This can be seen, for example, in Fig. 101, where some initial disturbance has created waves with a whole spectrum of wavelengths, and the long ones have travelled out further in a given time than the short ones.

Fig. 101 Dispersion of water waves.

Solitary waves

Now, in both the above small-amplitude theories the wave speed is *independent of the amplitude*, so neither can, in itself, explain Russell's experiments, where waves of greater height travel faster.

It was not until much later, in fact, that the whole phenomenon was given a mathematical explanation, most famously by Diederik Korteweg and Gustav de Vries, working in Amsterdam, in 1895. They showed that it was by a subtle combination of nonlinear and dispersive effects that the wave manages to maintain its form, and that this only happens if the wave has a very particular shape (Notes, p. 247).

They also deduced a simple expression for the wave speed:

$$c = \sqrt{gh}\left(1 + \frac{a}{2h}\right),$$

valid when the maximum height of the wave a above the undisturbed water level is much less than h. This reduces to Lagrange's formula when a is completely negligible compared with h, and is, in fact, consistent with an empirical formula suggested by Russell himself decades earlier.

In consequence, solitary waves of greater height a do indeed travel faster.

Yet it was not until the 1960s, with the rapidly developing computer revolution, that the Korteweg–de Vries equation was finally solved for the case when a large solitary wave chases and overtakes a small, slower-moving one (Fig. 102).

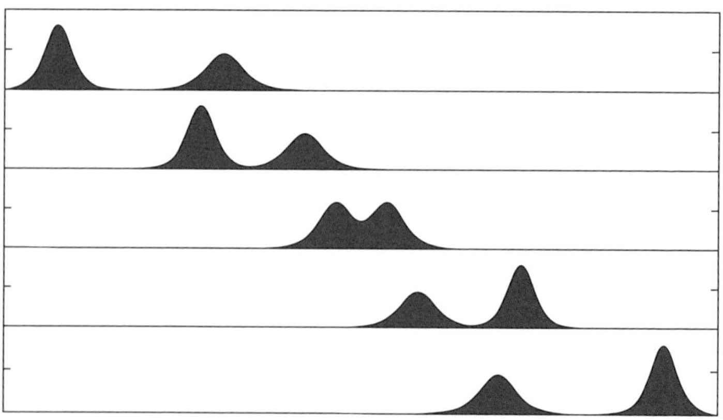

Fig. 102 A modern computation showing a large solitary wave overtaking a smaller one. (The time intervals between successive frames are equal.)

Most remarkably, both waves emerge from the interaction *completely unchanged in shape*, as Russell had, in fact, observed in his experiments, way back in 1837. And it is this extraordinary property, with the waves retaining their identities almost as if they were particles, that has given rise to the term 'soliton' in the modern literature.

And in all this there is just one telltale piece of evidence that something very unusual has happened, and that the two waves haven't simply passed through each other without any interaction at all.

For, as close inspection of Fig. 102 shows, the large-amplitude wave emerges from the interaction slightly ahead of where it would have been expected, and the small-amplitude wave is slightly 'late'.

27

SHOCK WAVES

One of the most familiar types of wave motion, surely, comes in the form of *sound*.

And because sound waves arise from the compressibility of a medium such as air, Euler's equations take on a rather different form, with the incompressibility equation replaced by one expressing conservation of mass.

In addition, the density ρ is no longer a given constant, but one of the variables of the problem, and an extra equation therefore appears, from thermodynamics.

* * * * * * *

An early theoretical attempt by Newton to deduce the speed of sound was based on an incorrect thermodynamic assumption, and it was eventually Pierre-Simon Laplace, in 1816, who first derived the correct formula:

$$c = \sqrt{\gamma p_0 / \rho_0}\,.$$

Here p_0 and ρ_0 denote the (constant) ambient pressure and density, while γ, with a typical value of 1.4 for air, denotes the ratio of specific heats.

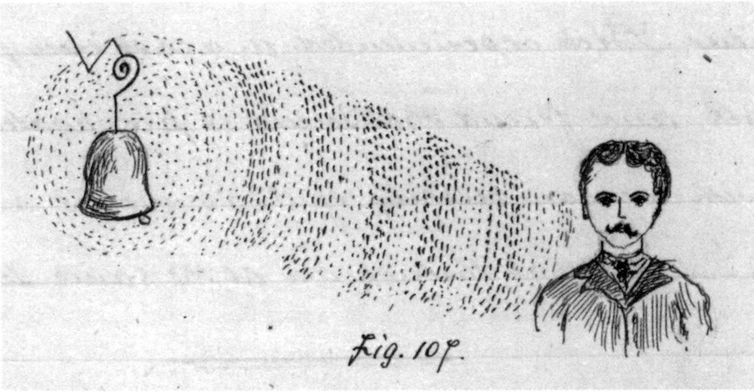

Fig. 103 Sound waves, from a young lady's handwritten physics exercise book in the author's possession, dated 1881.

The speed of sound is independent of wavelength, and sound waves are therefore nondispersive, like waves on a stretched string.

Unlike waves on a string, however, the fluid motions associated with sound waves are in the direction of travel, resulting in alternate regions of compression and expansion, as indicated schematically in Fig. 103.

* * * * * * *

The sound waves due to a *moving* source are of particular interest. And when an object moves at a constant speed v, the *Mach number* is defined in the well-known way as

$$M = v/c,$$

where c denotes the speed of sound.

Now, as the object moves a short distance past any particular fixed point P in space, it generates a sound wave which travels outward with speed c. So, at a time t later that particular disturbance is confined to a circle around P of radius ct.

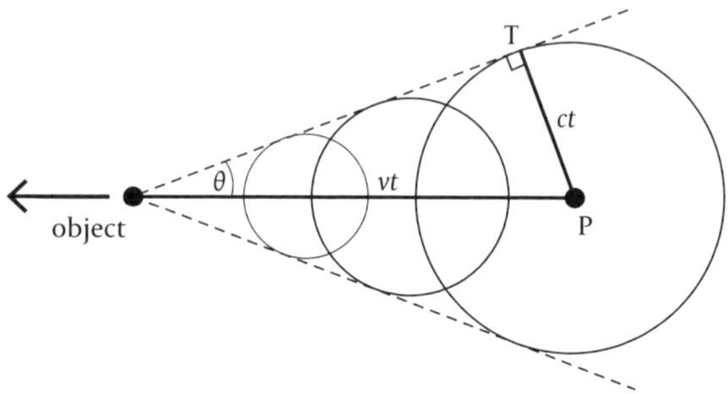

Fig. 104 Sound waves generated by a supersonically moving object.

By that time, the object itself will have moved on a distance vt, and will therefore be *outside* the circle in question if $M > 1$.

So, by the mechanism sketched in Fig. 104, the total acoustic disturbance caused by a *supersonically* moving object will be confined to a V-shaped wedge of half-angle θ, where

$$\sin \theta = 1/M,$$

and there will be no disturbance at all outside that region.

Shock waves

Now, all that we have seen so far has been based on the assumption that the sound waves in question have small *amplitude*, so that the changes in pressure and density are very small compared to their background or undisturbed values.

When this is not the case, everything changes, because it turns out that larger values of an acoustic disturbance then travel *faster than smaller ones.*

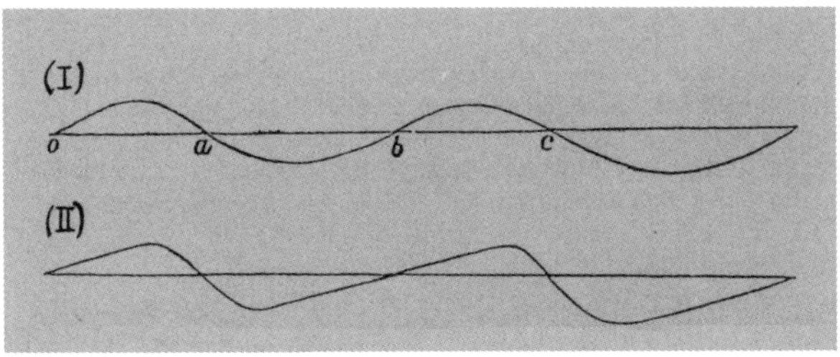

Fig. 105 Stokes's original sketches of 1848, illustrating the dependence of the fluid velocity u in a large-amplitude sound wave on x at two different times. (To emphasize the changing *shape* of the waveform, a uniform displacement to the right at the small-amplitude sound speed c has been omitted.)

By 1848, in a now-famous paper entitled 'On a Difficulty in the Theory of Sound', Stokes had noticed that this causes a serious problem, remarking with reference to Fig. 105 that

> it is evident that in the neighbourhood of the points a, c the curve becomes more and more steep as t increases.

And this continual steepening leads eventually to $\partial u/\partial x$ becoming *infinite* at certain points *in a finite time*, indicating a complete breakdown of the whole theory.

Stokes then suggested, rather tentatively, what might actually happen instead:

> Perhaps the most natural supposition to make for trial is, that a surface of discontinuity is formed, in passing across which there is an abrupt change of density and velocity.

This idea of a *shock wave* was wholly new at the time, and not readily accepted. Even 30 years later, in fact, Lord

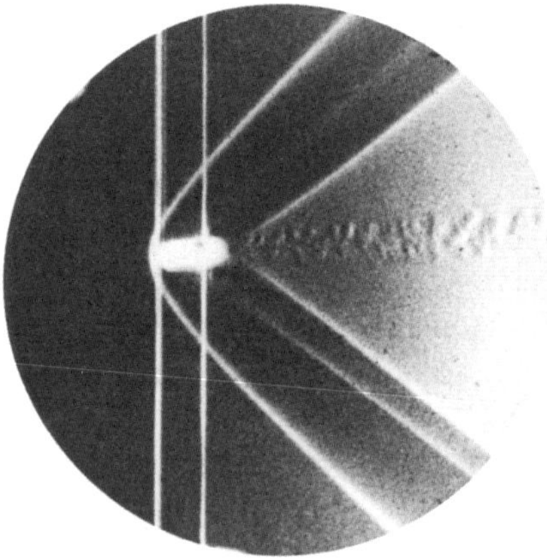

Fig. 106 Photograph by Mach of the shock wave caused by a speeding bullet. The two vertical lines are part of the 'Schlieren' apparatus used to capture the image.

Rayleigh wrote to Stokes raising an objection to the idea, and Stokes replied most diffidently, adding

> P.S. You will observe I wrote somewhat doubtfully about the possibility of the queer motion.

Yet, by 1887, experiments by Ernst Mach in Germany had actually visualized shock waves, most famously in the photograph shown in Fig. 106.

And from a purely mathematical point of view, the kind of wave-steepening shown in Fig. 105, and the problems which that can create, continue, to this day, to be one of the most characteristic features of nonlinear dynamics of all kinds.

28

VORTEX MOTION

Another area of fluid mechanics where nonviscous theory often works well is vortex motion.

An individual vortex typically consists of a small and rapidly spinning *core*, surrounded by a spinning motion which diminishes with distance from the core itself. And by an elegant theorem due to Hermann von Helmholtz in 1858, a vortex gets swept about at the local flow velocity due to everything else.

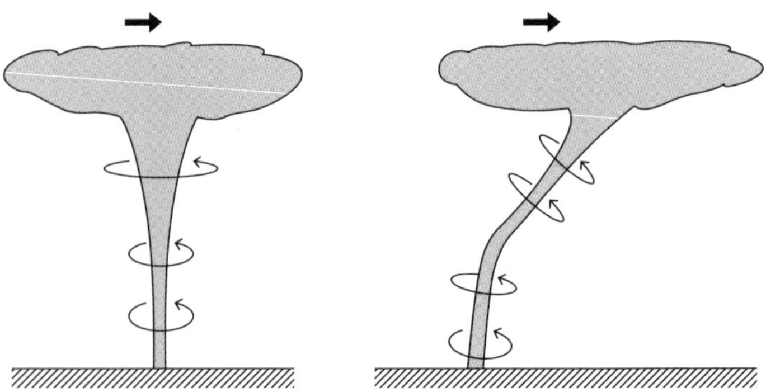

(a) (b)

Fig. 107 Vortex motion in action.

The tornado in Fig. 107 provides a good example, for as the thundercloud above moves on, it tilts the vortex itself, in accord with Helmholtz's theorem.

Vortex rings

A rather more intriguing situation arises if we consider not a straight vortex but a *ring*-shaped one, for it turns out that this can propel *itself* along, essentially by the same mechanism.

In January 1867, for instance, Lord Kelvin wrote to Helmholtz in Germany about some experiments on vortex motion that he had just been shown in Edinburgh:

> Take one side (or the lid) off a box (any old packing case will serve) and cut a large hole in the opposite side. Stop

Fig. 108 Vortex rings, made visible by smoke. (From John Perry's *Spinning Tops* of 1929.)

> the open side loosely with a piece of cloth, and strike the middle of the cloth with your hand. If you leave anything smoking in the box, you will see a magnificent ring shot out by every blow.

For a smoke ring is, essentially, a ring-shaped *vortex*, and Helmholtz had predicted exactly this behaviour in his 1858 paper.

'Leapfrogging' of vortex rings

Now, because the flow speed outside the narrow vortex core decreases with distance, a vortex ring of large radius sweeps itself along more slowly than a small one. And a particularly interesting situation then arises when one such ring chases another (Fig. 109).

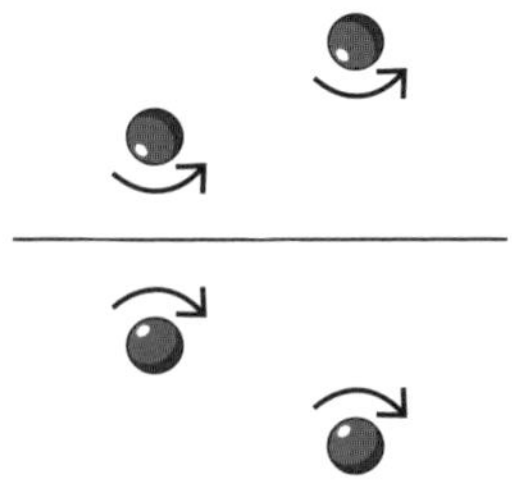

Fig. 109 One vortex ring chasing another.

For, not only will the smaller, chasing ring be moving along faster, but the flow generated by its vortex core will tend to 'open up' the ring in front and therefore slow it down. And, by exactly the same mechanism, the flow due to the one in front will tend to reduce the radius of the one behind, making it proceed even faster.

In this way, then, it can quite easily happen that one ring *passes through* another, and the sequence of events can then begin all over again.

In other words, the two vortex rings can, in principle, *repeatedly* change places in a sort of 'leapfrogging' motion.

A doomed atomic theory

In his letter to Helmholtz of 1867, Kelvin is rather more circumspect, simply remarking that he can 'sometimes' make one ring 'shoot through' another. And, in any case,

Fig. 110 Kelvin's sketches of knotted and linked vortex rings, for his 'vortex atom' theory of matter.

it becomes quite clear later in the same letter that his real interest in vortex rings is rather more exotic:

> The absolute permanence of the rotation . . . shows that if there is a perfect fluid all through space, constituting the substance of all matter, a vortex-ring would be as permanent as the solid hard atoms assumed by Lucretius and his followers.

This extraordinary (but doomed) atomic theory, 40 years ahead of that of Niels Bohr, seems to have been the main driving force behind Kelvin's fluid mechanics research at the time. It was the sheer *permanence* of a vortex ring, apparently, that most fired his imagination.

And my mind therefore rather boggles as to what Lord Kelvin would have made of a curious experiment that I once saw performed at a research meeting at University College, London . . .

A new twist on an old problem

The lecturer took a large bowl of water, and immersed two syringes in it, which he had adapted to produce liquid vortex rings—one dyed red and the other dyed blue.

He arranged an angle of about 30° between the two syringes, and then fired off a red ring and a blue ring simultaneously so that they would travel side by side in a horizontal plane and collide. The rings interacted in the

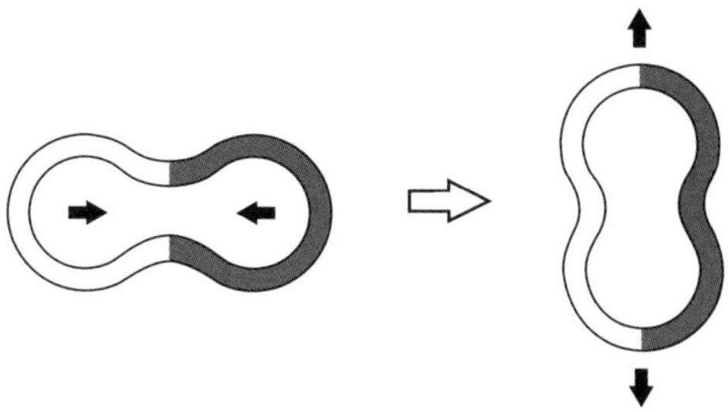

Fig. 111 The brief interaction between two vortex rings during a collision in water. The small arrows give a rough indication of how the rings are travelling before and after the interaction, though they are also moving towards us throughout.

manner sketched roughly in Fig. 111, availing themselves of the fact that water is slightly viscous, and then went their separate ways again.

But this time they were moving in a vertical plane, and *each ring was one-half red and one-half blue*!

29

THE CURSE OF THE GOLDEN RATIO

One of the earliest theoretical investigations of vortex leapfrogging, by the mathematician Augustus Love in 1894, involved not circular rings but two identical vortex *pairs*.

A vortex pair consists of two straight vortices of equal strength but opposite sign (Fig. 112). Each then sweeps the other along, so that they travel together in much the same way as a vortex ring.

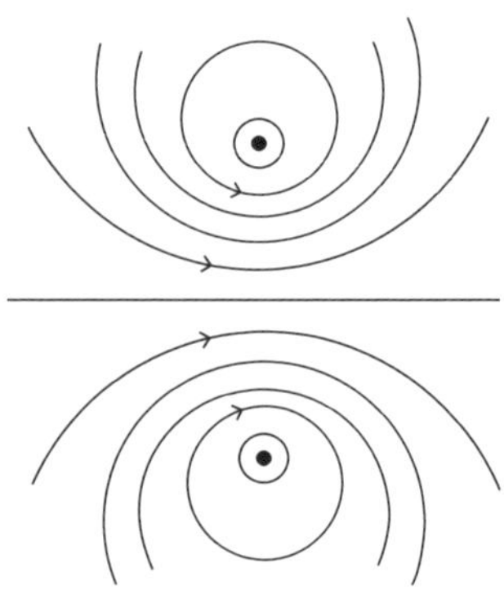

Fig. 112 The flow due to a vortex pair.

And when one pair chases another, leapfrogging can again occur, though Love found that the nature of the motion then depends critically on r, the ratio of the breadths of the two pairs at the instant when one passes through the other.

In particular, Love proved that leapfrogging is only possible if

$$r > 3 - 2\sqrt{2}.$$

(a) (b)

Fig. 113 (a) The breadth ratio r. (b) An impromptu sketch of Augustus Love (1863–1940) towards the end of his life, by an undergraduate at Oxford.

If r is less than this critical value of 0.172, the inner pair passes through the outer pair so quickly that the outer, slower-travelling pair is unable to catch up, and the two just move steadily further apart as time goes on.

And so the problem remained, until, by chance, I took it up again, over 100 years later.

* * * * * * *

With the aid of a computer to solve the differential equations, I soon confirmed that leapfrogging is indeed possible when r exceeds Love's critical value of 0.172, but found that it is then typically *unstable*, so that any small initial disturbance causes the original periodic motion to quickly break up.

Moreover, for r bigger than about 0.29, this break-up often happens in a most bizarre way—the vortices typically continue in a disorderly leapfrogging motion punctuated by periods of three-vortex 'walkabout' (Fig. 114).

What seems to happen is that two like-signed vortices revolve rapidly around one another, and sweep one of the other, oppositely signed, vortices out of its usual path and into a roughly circular one instead. Walkabout episodes of this kind take place on seemingly unpredictable sides of the main path and at seemingly irregular intervals.

And, according to my computer solutions of the differential equations, Love's original leapfrogging motion only becomes stable when

$$r > 0.382,$$

i.e. roughly twice the value at which leapfrogging becomes possible at all.

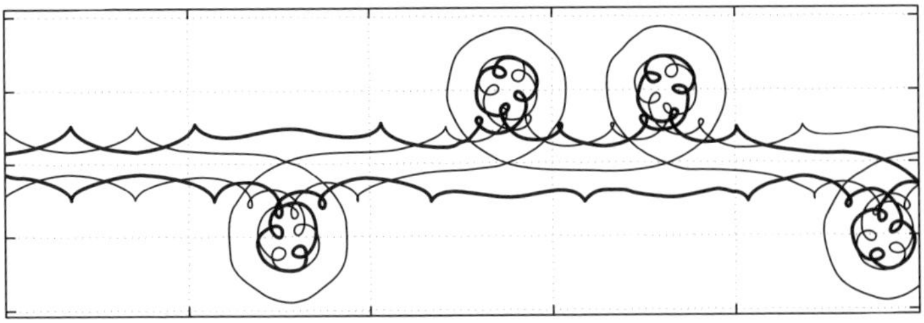

Fig. 114 'Walkabout' instability of vortex pair leapfrogging, with $r = 0.31$.

I published all this in 2000, and after causing a fair amount of interest at the time the whole problem died down again.

* * * * * * *

Then, in 2012, I received an email, quite out of the blue, from Laust Tophøj, a physicist from Denmark, about some work he had been doing with the vortex motion expert Hassan Aref.

They had conducted a much more sophisticated mathematical analysis of the whole problem, confirming that 0.382 is indeed the approximate value of r at which the leapfrogging becomes stable, but showing that the *exact* critical value of r for stability is

$$r = 1/\phi^2,$$

where ϕ is the golden ratio!

* * * * * * *

Now, as many readers will know, the golden ratio ϕ has a long history in mathematics (Fig. 115).

It started life in ancient Greece as the ratio of diagonal to side for the regular pentagon, and then made a famous reappearance in 1202 in connection with the Fibonacci sequence

$$1,\ 1,\ 2,\ 3,\ 5,\ 8,\ 13,\ 21\ldots,$$

where each term (after the second) is the sum of the previous two. For the ratio of successive terms gets closer and closer to ϕ as the sequence goes on.

And if by any chance you are wondering what on earth the golden ratio is doing in fluid mechanics, I wondered the same in 2012. But it was, in fact, even worse than that.

$$\phi = \frac{1+\sqrt{5}}{2} = 1.618\ldots$$

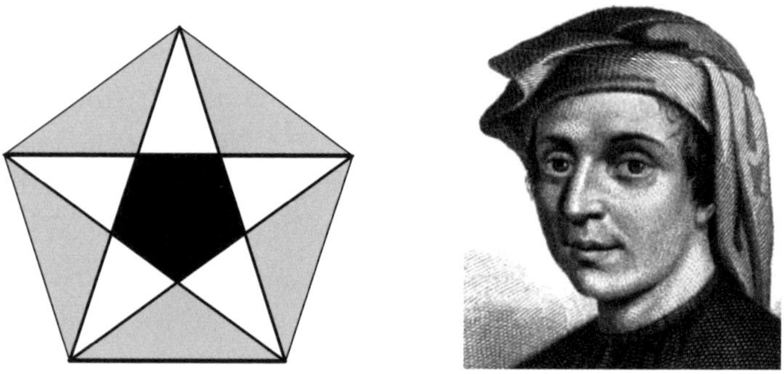

Fig. 115 The golden ratio, the regular pentagon, and Fibonacci (c.1170–1250).

For once we get beyond the regular pentagon and the Fibonacci sequence, the golden ratio is a somewhat divisive number in mathematics. There are those who see it almost everywhere—in nature, art, and even music—yet there are many others who view most of all that as wishful thinking.

And, in truth, I had always been a firm member of the second camp. Occasionally, even, I would express deep cynicism about the golden ratio quite openly in my public lectures on mathematics.

Yet here it was, turning up, quite out of the blue, in my own research on vortex motion.

And I hadn't even noticed.

I should really have known better, if only because, in my opinion at least, one of the great attractions of mathematics at its best lies in the many surprising *connections* between apparently quite different parts of the subject.

30

THE CORIOLIS FORCE

One of the strangest pendulums in history appeared at the Pantheon in Paris, in 1851.

It was designed by the physicist Léon Foucault, and consisted of an enormous lead bob suspended from the central dome by a wire so long that each complete oscillation took about 16 seconds (Fig. 116).

Fig. 116 The Foucault pendulum.

In addition, a pin was attached to the underside of the bob which made tracks in a small ridge of sand on the surrounding circular rail. And as the pendulum swung to and fro, the plane of oscillation *rotated very slowly clockwise* (when viewed from above), so as to complete one rotation in about 32 hours.

And the point of it all was to provide proof of the Earth's rotation . . .

The rotating drum revisited

To understand something of this, recall first the people on the wall of the large rotating drum in Fig. 1.

We noted in Chapter 4 that, because of their circular motion, they are accelerating all the time towards the rotation axis. Moreover, this acceleration is caused by a radially inward force on each person exerted by the wall of the drum.

Now, this is indeed the correct explanation, according to the laws of dynamics, from the viewpoint of a fixed observer, watching the drum as it goes round.

Yet, quite reasonably, the people themselves view things very differently. As far as they are concerned, they are *at rest*, and the reason they don't move is that—as they see it—the inward force due to the drum is exactly balanced by a somewhat mysterious *centrifugal*, or radially *outward* force which is pinning them to the wall.

This is a clear indication, then, that if we wish to use our normal ideas of mechanics in a rotating frame of reference—essentially as if that frame were *not* rotating—then we have to include additional 'apparent' forces due to the frame rotation.

And as soon as one of the people in Fig. 1 starts *moving*, relative to the rotating frame of reference, an even greater subtlety comes into play...

The Coriolis force

Imagine, for instance, that the floor of the drum is in place, and that it is frictionless. And imagine, too, that the man has finally had enough of being pinned to the wall, and with superhuman effort propels himself towards the rotation axis. What will happen?

Well, from the viewpoint of a stationary observer watching the drum go round, the answer is clear: as he moves radially inward he will have to conserve his angular momentum about the rotation axis, so his angular velocity about that axis will therefore increase, rather like the ice-skater in Fig. 55.

In the rotating frame, then, as the man propels himself inward, he will find his path being *turned*, as if by yet another mysterious force.

The 'apparent' force in question is called the *Coriolis force*, and in the rotating frame of Fig. 117 it acts to deflect *any*

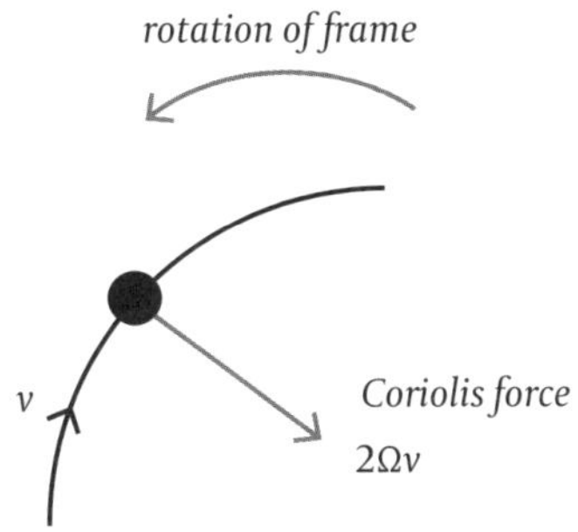

Fig. 117 The Coriolis force (per unit mass) on an object moving in a rotating frame pf reference.

relative motion to the right. Its magnitude (per unit mass) is $2\Omega v$, where Ω is the angular velocity of the frame and v the speed of the relative motion.

The Foucault pendulum

At this point we can finally make sense of the strange pendulum in Fig. 116.

The corresponding formula for the Coriolis force in that case is

$$\textit{Coriolis force/unit mass} = 2\Omega v \sin\phi,$$

where v denotes the velocity of the object in question relative to the Earth, Ω denotes the angular velocity of the Earth itself, and ϕ denotes latitude. The Coriolis force deflects relative motions to the right in the northern hemisphere, where ϕ is positive, and to the left in the southern hemisphere.

In Paris, therefore, it turns each swing of the pendulum very slightly to the right, and this is what leads to a very slow rotation of the oscillation plane. This rotation is clockwise (when viewed from above), and the time for one whole revolution turns out to be

$$\frac{2\pi}{\Omega \sin\phi},$$

which, at the latitude of Paris, works out at about 32 hours.

* * * * * * *

While Foucault's experiment caused something of a sensation at the time, we will see later how the real significance of the Coriolis force became apparent only rather later, with the rise of meteorology, oceanography, and even astrophysical fluid mechanics, where it can often play an almost dominant role.

31

THE THREE-BODY PROBLEM

Throughout the eighteenth and nineteenth centuries much consternation was caused by one particular problem, in which three point masses all attract one another according to the inverse-square law of gravitation.

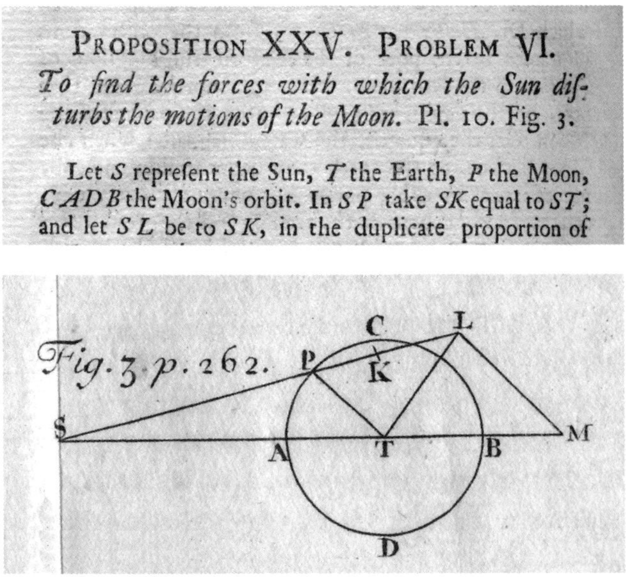

Fig. 118 From the first English edition of Newton's *Principia* (1729).

The problem originated, like so many others, with Isaac Newton, who was trying to account for the motion of the Moon under the influence of both the Sun and the Earth (Fig. 118). And this was evidently a real struggle, for according to a contemporary, it was the one problem that 'made his head ache'.

* * * * * * *

In due course, others took up the challenge, and one elegant three-body solution was discovered by Lagrange in 1772, where all three rotate in ellipses about a common focus.

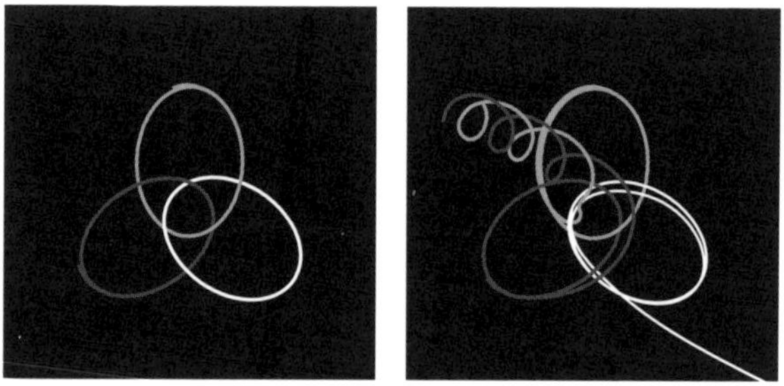

(a) (b)

Fig. 119 (a) A three-body solution by Lagrange, and (b) its instability.

A particular example, with three equal masses, is shown in Fig. 119a, though the motion is, in fact, unstable. For, if one of the three bodies starts very slightly out of place, as in Fig. 119b, everything soon evolves quite differently.

* * * * * * *

A major mathematical development came in 1835, when the great Irish mathematician William Rowan Hamilton (1805–1865) took Lagrange's equations of mechanics, and reformulated that particular approach to the subject yet again, putting the fundamental equations into the form

$$\frac{dq_i}{dt} = \frac{\partial H}{\partial p_i},$$
$$\frac{dp_i}{dt} = -\frac{\partial H}{\partial q_i}.$$

Here, the q_i again denote 'generalized coordinates', while the p_i are 'generalized momenta', and H denotes the so-called Hamiltonian, which in many cases of practical interest (though certainly not all) is the total energy of the system.

This new approach made it possible to view the evolution of a dynamical system not in the usual physical space but in a typically much higher-dimensional (and wholly mathematical) 'phase space', where the whole state of the system in question at any one time can be represented *by a single point*.

In some ways, the use of phase space can turn dynamical questions into essentially *geometric* ones instead. One natural question to ask of any complex system, for instance, is 'Are there any *periodic* solutions?', and this is equivalent to asking whether there are any *closed* curves in phase space.

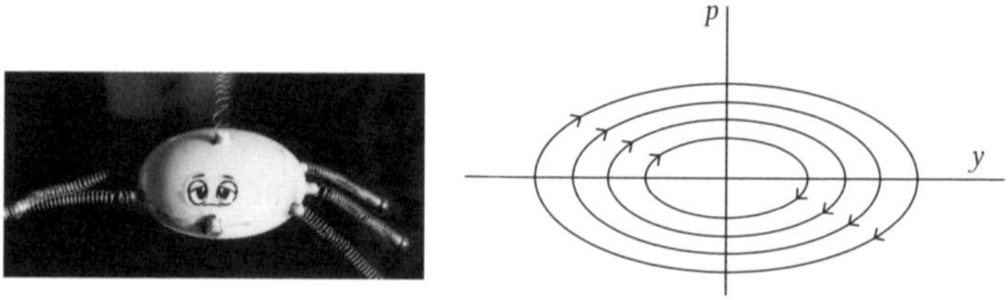

(a) (b)

Fig. 120 (a) The oscillating spider and (b) paths in phase space.

In the simple case of the toy spider of Chapter 12, for instance, the state of the system at any moment can be described by just the spider's vertical displacement y and its upwards momentum $p = mv$. This yields a two-dimensional phase space, as indicated in Fig. 120b.

And in this particular case, *all* the phase paths are closed, because no matter how we start the motion the spider just bounces up and down in an endlessly repetitive way.

* * * * * * *

This whole new approach was turned to dramatic effect in 1887, when King Oscar II of Sweden offered a prize of 2500 crowns for an essay on the question 'Is the Solar System Stable?' The prize was awarded to Henri Poincaré, who

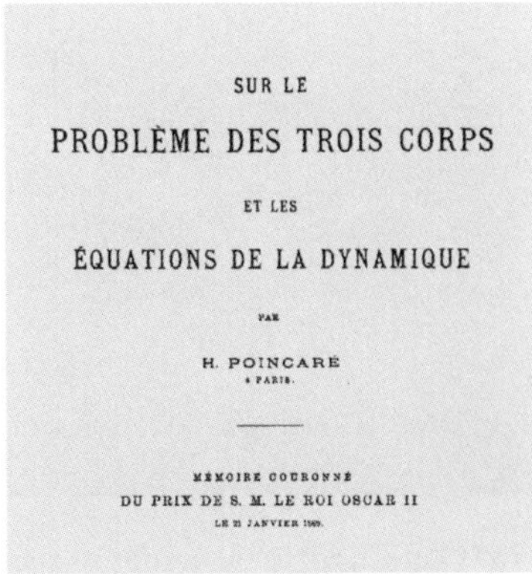

SUR LE

PROBLÈME DES TROIS CORPS

ET LES

ÉQUATIONS DE LA DYNAMIQUE

PAR

H. POINCARÉ

À PARIS.

MÉMOIRE COURONNÉ

DU PRIX DE S. M. LE ROI OSCAR II

LE 21 JANVIER 1889.

Fig. 121 Henri Poincaré (1854–1912) and his essay of 1889 on the three-body problem.

focused on the so-called 'restricted' three-body problem, in which the third mass is so small that it is essentially passive, and has no gravitational effect on the first two.

Yet even in this case Poincaré wrote of some of the paths in phase space:

> One is struck by the complexity of this figure which I am not even attempting to draw. Nothing can give us a better idea of the complexity of the three-body problem.

And it is now recognized that Poincaré had glimpsed in this way, as early as 1889, what we now call *chaos*.

* * * * * * *

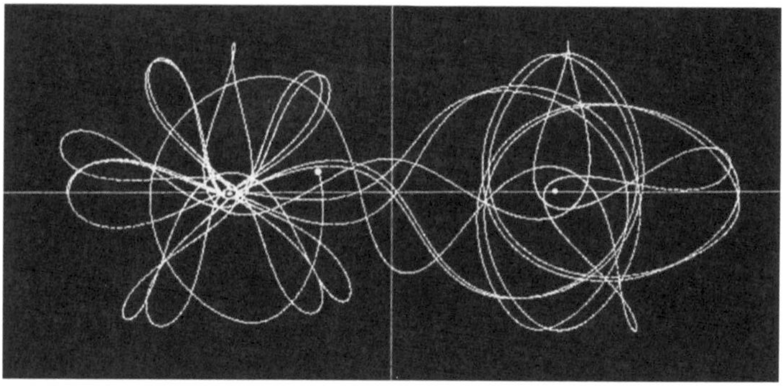

Fig. 122 Chaotic shuttling in the restricted three-body problem.

Today, with a modern laptop, we can generate such chaotic solutions fairly easily, and Fig. 122 shows one example, where two large and equal masses rotate in a circle about their mid-point while a tiny third mass is shuttled to and fro between them. The viewpoint in Fig. 122 is from a frame which rotates with the two primary masses, and the shuttling is seen to take place in a seemingly haphazard way.

32

A PARADOX RESOLVED

In August 1904, at the Third International Congress of Mathematicians in Heidelberg, Ludwig Prandtl presented a paper 'On the Motion of Fluids of Very Small Viscosity'. And in a lecture of just ten minutes he outlined the solution to one of the most outstanding problems in fluid mechanics.

Fig. 123 Ludwig Prandtl (1875–1953) at his experimental water-tank.

For while Euler's nonviscous flow theory worked very well on problems of wave motion or vortex dynamics, it had been known for 150 years that it typically gave completely the wrong answer for flow past an object, *even when viscosity appeared to be negligible.*

The most well-known manifestation of this was d'Alembert's paradox, as described in Chapter 24, but the result of zero drag was, in itself, really no more remarkable than the way in which the whole flow was wrong.

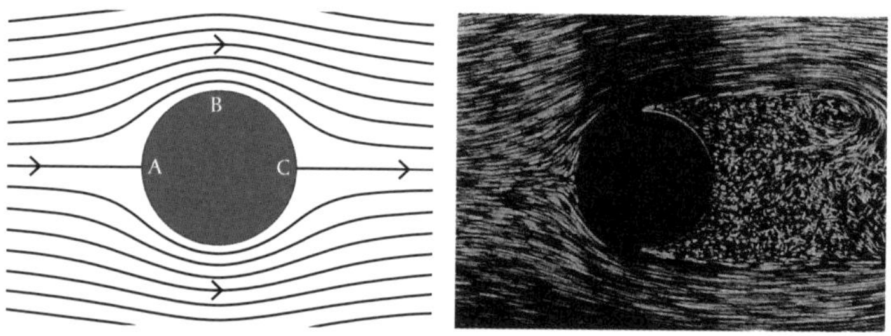

(a) (b)

Fig. 124 Flow past a circular cylinder (a) according to nonviscous theory and (b) in practice, at a Reynolds number of 2000 based on the cylinder diameter.

In Fig. 124a, for instance, we see flow past a circular cylinder according to the nonviscous theory, with streamlines which are neatly symmetric fore and aft. The real, observed flow of a fluid of small viscosity in Fig. 124b, however, is completely different, with a large 'wake' behind the cylinder.

And Prandtl finally provided the explanation.

Viscous boundary layers

In the nonviscous theory, fluids *slip* freely along solid surfaces like the cylinder in Fig. 124. Yet, in reality, there is no such slip, even with fluids—like water—of very small viscosity.

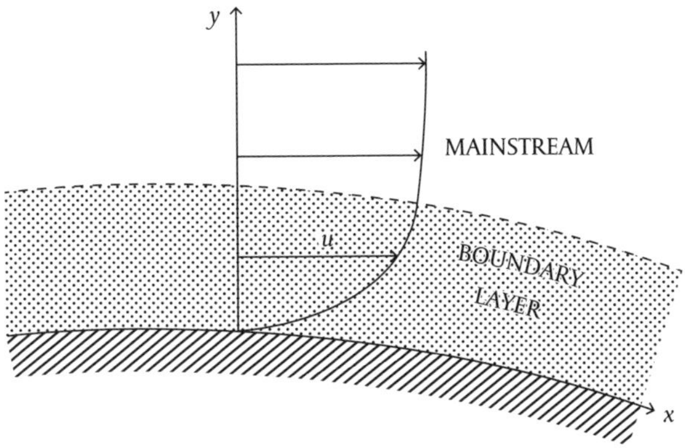

Fig. 125 The boundary layer.

And Prandtl showed that for flow at high Reynolds number there is, instead, a very thin *boundary layer* which rapidly adjusts the mainstream flow speed to satisfy a 'no-slip' condition on the cylinder itself (Fig. 125). Moreover, because this layer is so thin, the velocity varies so rapidly across it that viscous stresses become important there, even if the coefficient of viscosity μ itself is very small.

Prandtl showed, in fact, that the boundary layer thickness is proportional to $\sqrt{\mu}$, so that the smaller the viscosity μ, the thinner the boundary layer, and the more intense the shearing flow within it.

Viscous boundary layers can even be seen quite clearly in Fig. 126, where water flows into a narrowing channel, the flow being visualized by hydrogen bubbles, which act a bit like dye and mark individual fluid elements.

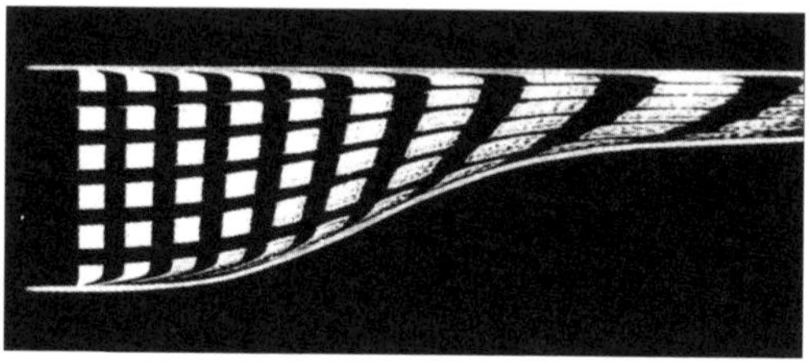

Fig. 126 Flow of water into a narrowing channel.

Separation

In flow past a circular cylinder, however, the consequences are more profound.

For the nonviscous theory predicts a slip velocity at the rear which falls from a maximum at the point B in Fig. 124 to zero at the rear stagnation point C. By Bernoulli's theorem, then, there is very substantial *increase* in pressure as we proceed from B to C, and Prandtl showed that this adverse pressure gradient impresses itself on the whole of the flow in the viscous boundary layer.

And this adverse pressure gradient actually drives the slowest-moving fluid in the boundary layer *backwards*, which causes the whole layer to *separate* from the boundary (Fig. 127).

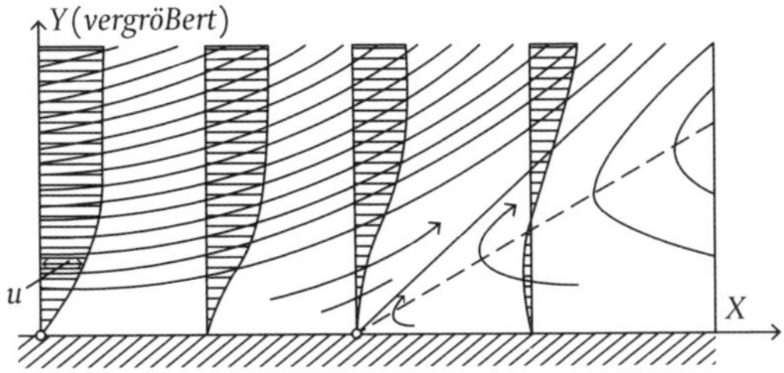

Fig. 127 The original sketch of viscous boundary layer separation in Prandtl's paper of 1905.

This in turn completely changes the bulk of the flow at the rear of the body. And in this kind of way, even with a very small coefficient of viscosity μ, the strange paradox noticed by d'Alembert in 1752 was at last resolved.

Dynamics at tea-time

Separation is not, however, the only way in which thin viscous boundary layers can effectively control an entire flow.

The whole process by which a stirred cup of tea comes to rest, for instance, is much more subtle than one might imagine, and entirely controlled by a thin viscous boundary layer on the bottom of the cup.

What happens is this. At first, the non-viscous main body of the fluid is rotating rapidly, and a radially inward pressure gradient keeps the fluid elements on their roughly circular paths.

This inward pressure gradient then impresses itself on the more slowly rotating fluid in the bottom viscous boundary layer, causing it to spiral *inward*. Needing somewhere to go, that fluid then turns upward into the nonviscous main body of the fluid, *and then out again* to complete the secondary circulation indicated in Fig. 128.

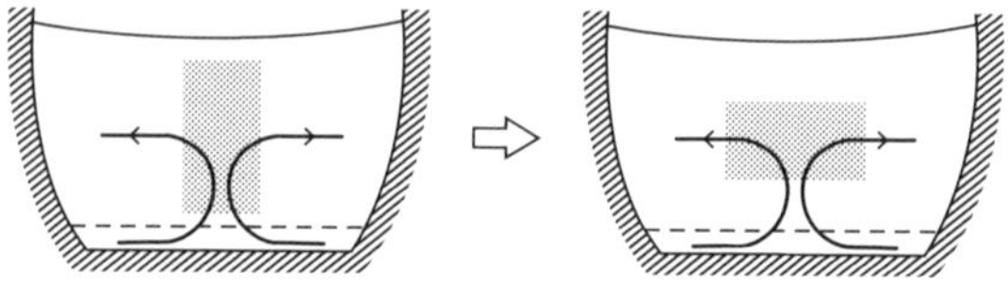

Fig. 128 The secondary circulation in a stirred cup of tea.

And this secondary circulation turns what was a tall, thin, rapidly rotating blob in the main body of the fluid into a shorter, fatter one, which consequently spins more slowly than before, through conservation of angular momentum.

The whole spin-down mechanism is therefore a subtle mix of nonviscous fluid mechanics in the main body of the fluid and viscous effects in the bottom boundary layer.

And just in case you should find this all a bit hard to believe, the evidence is actually in plain sight, because it is the inward motion in the bottom boundary layer that accounts for the well-known way in which tea leaves gather, during the whole spin-down process, *in the middle* on the bottom of the cup.

33

THE SECRETS OF FLIGHT

The way in which a wing acquires lift can be a curiously emotive topic.

Even today, popular articles in newspapers and magazines occasionally take the line: 'Why do mathematicians make the whole business of lift on a wing so complicated? It's all quite simple, really.'

So far as I can determine, however, it *isn't*.

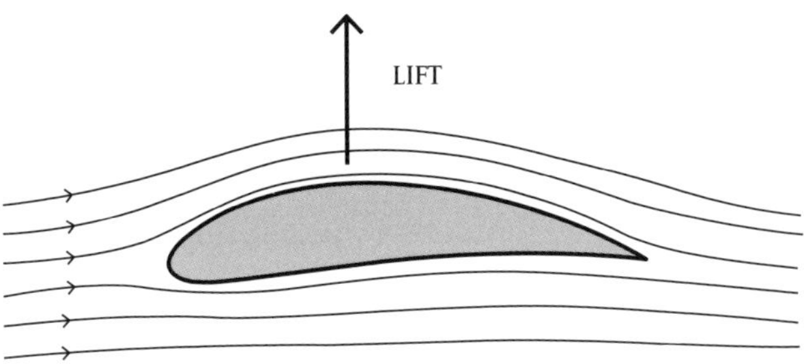

Fig. 129 Flow past a wing.

Nonetheless, the short version of the correct explanation comes in just two parts.

First, note that all the streamlines in Fig. 129 originate far upstream, where the oncoming flow speed and background pressure are both constant. Bernoulli's theorem from Chapter 24 then tells us that

$$p + \frac{1}{2}\rho v^2 = \mathit{constant}$$

everywhere, where p denotes pressure associated with the fluid motion, ρ air density, and v flow speed.

Second, the air flows faster over the top of the wing than it does along the bottom. The downward pressure on the top of the wing is therefore *less* than the upward pressure on the bottom, giving rise to lift.

Yet this leaves, of course, one major question . . .

Why is the airflow faster over the top of the wing?

One slightly absurd explanation for this is that, because of the shape of the wing, a small fluid element moving over the top has 'further to go'.

Insofar as this argument makes any sense at all, it seems to assume that the two journeys take the same time, with two small fluid elements which part at the front of the wing meeting up again at the trailing edge. But there's no earthly reason why they should meet up again like this, and experiment shows that they don't (Fig. 130).

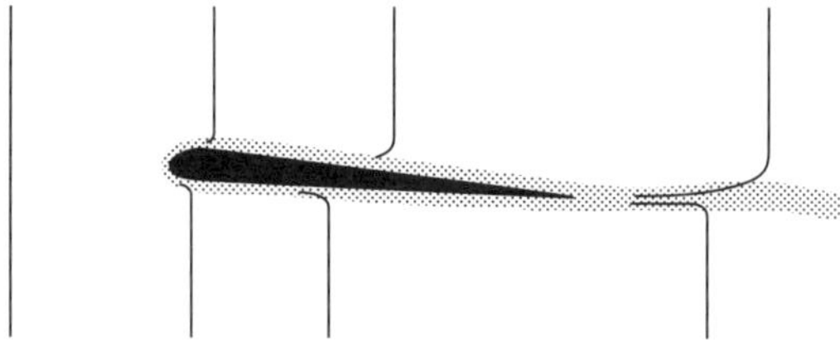

Fig. 130 The fate of a marked line of fluid in flow past a wing.

The real explanation seems to have been first grasped by the engineer Frederick Lanchester.

At a meeting of the Birmingham Natural History and Philosophical Society in 1894, he presented a paper on 'The Soaring of Birds and the Possibilities of Mechanical Flight'. It contained many of the key elements of the correct explanation of lift, but not in conventional scientific terms, and it is doubtful if anyone understood it.

And when Lanchester eventually submitted his paper for publication by the Physical Society, it was rejected.

Much later in life, in 1931, he recalled those early days, and wrote

> So far as aeronautical science is concerned, I cannot say that I experienced anything but discouragement . . . and I was seriously warned that my profession as an engineer would suffer if I dabbled in a subject that was merely a dream of madmen!

Yet, by then, many of Lanchester's ideas had been proved correct, largely through the work of Martin Kutta in

Fig. 131 F. W. Lanchester (1868–1946), in his garden in Birmingham.

Germany and Nikolai Joukowski in Russia. And Joukowski, in fact, discovered a simple formula for the lift (per unit length of span):

$$Lift = \rho U \Gamma,$$

where ρ is the fluid density, U the upstream flow speed, and Γ the so-called 'circulation' round the wing.

The circulation theory of lift

The precise definition of circulation is a little technical (Notes, p. 251), though it has dimensions of *speed* × *length*. With reference to Fig. 129 it can be thought of loosely as characterizing a clockwise circulatory flow around the wing itself, which therefore augments the flow over the top of the wing but opposes that below. And, in conjunction with Bernoulli's theorem, this explains the lift.

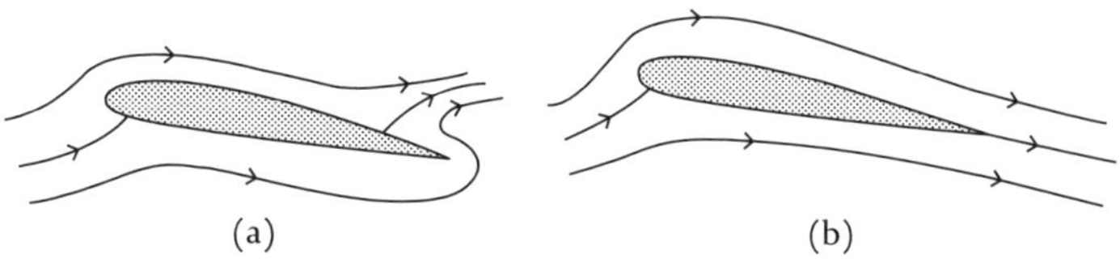

Fig. 132 Nonviscous flow past a wing with (a) zero circulation and (b) sufficient clockwise circulation to give smooth flow at the trailing edge.

But where does this clockwise circulation come from?

The answer to this is subtle, and was only fully worked out by Prandtl, in 1923.

When a wing is suddenly started into motion, the flow past it initially has *zero* circulation, and the streamlines have the general form shown in Fig. 132a, with a stagnation point on the upper surface near the trailing edge.

As the lower fluid desperately tries to move round the sharp trailing edge, the thin viscous boundary layer on the wing separates, shedding a so-called *starting vortex* (Fig. 133). As this vortex spins anticlockwise there is anticlockwise circulation round it, and it is in reaction to this that an equal but opposite *clockwise* circulation is gradually established round the wing itself (Notes, p. 251). This whole process then continues until the circulation round

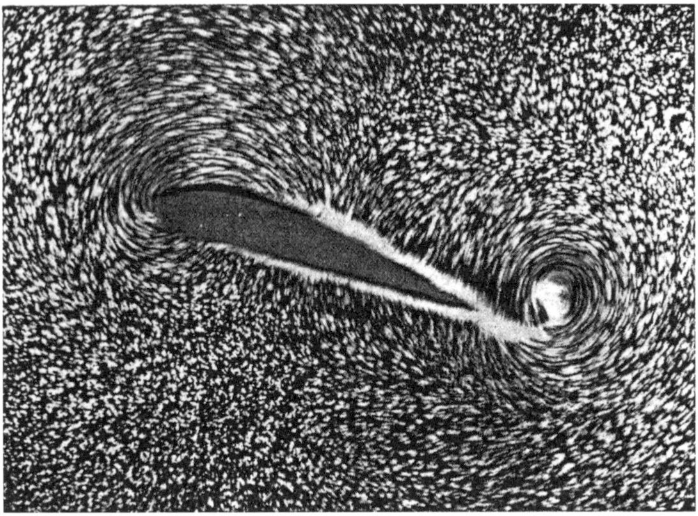

Fig. 133 The starting vortex.

the wing is sufficient to allow the flow to leave the trailing edge smoothly, as in Fig. 132b.

In this way, then, it is the starting vortex in Fig. 133 that is ultimately responsible for the clockwise circulation round the wing, and therefore the lift force on it.

A new twist on an old problem

While all this was more or less understood by the 1930s, an exotic variation on the whole theme was discovered in 1973 by Torkel Weis-Fogh in the hovering motions of the tiny chalcid wasp *Encarsia formosa*, with a wing chord of just 0.2 mm.

For this tiny insect generates its lift by *clapping its wings together*, and then flinging them open about a horizontal line of contact, so that air has to rush in to fill the gap (Fig. 134).

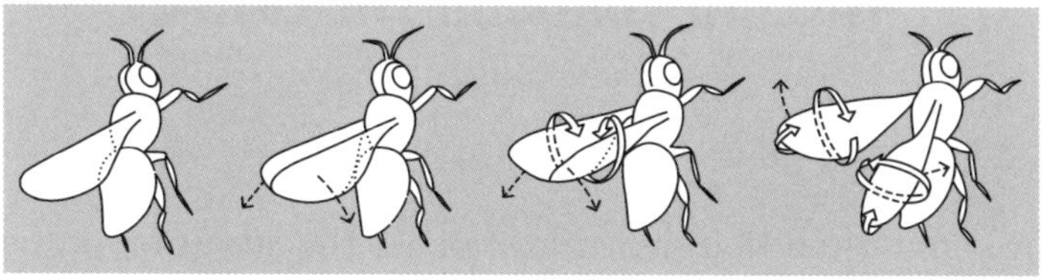

Fig. 134 The Weis-Fogh clap-and-fling lift mechanism.

As a result, when it moves its wings apart again, each has circulation about it of the correct sign to give lift. Most unusually, then, each wing essentially acts as the 'starting vortex' for the other.

And the whole novel mechanism is all the more impressive for the fact that *Encarsia formosa* goes through the entire routine in Fig. 134 roughly 400 times a second.

34

FALLING THROUGH THE EARTH

It is not often that dynamics catches the attention of the general public, but the French writer Camille Flammarion certainly managed it in 1909 with an unashamedly fanciful article in the *Strand Magazine*.

For imagine, if you will, that you are falling down a large hole that goes *right through the centre of the Earth* (Fig. 135).

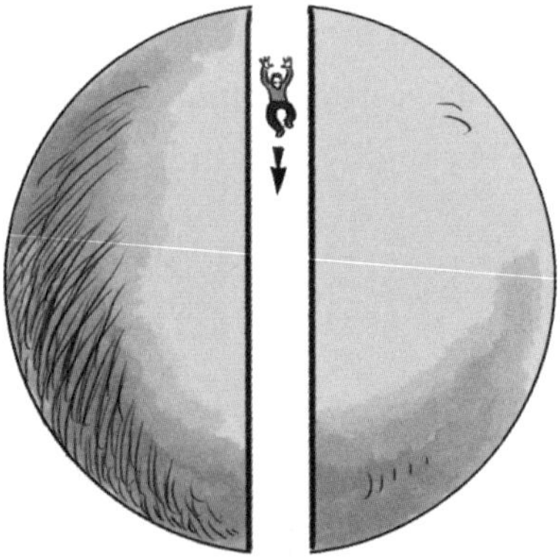

Fig. 135 Falling through the Earth.

Then, if the Earth were simply a uniform solid sphere, and all physical effects other than gravity could be neglected, you would emerge at the other end after a

$$journey\ time = \pi\sqrt{R/g},$$

where R is the radius of the Earth and g the acceleration due to gravity at the Earth's surface. And this works out at just over 42 minutes.

More remarkably still, this gravitational transit time would be the same *for all other straight tunnels through the Earth*, provided only that they start and end somewhere on the Earth's surface.

Fig. 136 Flammarion's idea of a hole through the Earth as a transport system (1909).

It wouldn't matter, then, if you were going from London to Sydney or from London to Oxford; the free ride, courtesy of gravity alone, would still take 42 minutes.

* * * * * * *

The special case when the hole passes through the centre of the Earth has, in fact, a much longer history.

Galileo, for instance, in his *Dialogue Concerning the Two Chief World Systems* (1632), writes:

> if the earth were tunnelled through the centre, and the ball were let fall . . . toward the centre, I verily believe that it would pass beyond the centre and ascend so much as it had descended.

And even Euler, roughly a century later, considers the problem briefly and claims that the object will oscillate about the centre of the Earth.

The first detailed analysis, however, seems to have been in *The Ladies' Diary* for 1781. This was a popular journal, published annually in London, which ran from 1704 to 1841. It was, in part, an almanac, but it also contained 'Delightful and Entertaining Particulars', which included mathematical problems submitted by its readers.

And in the 1781 edition, we find, as 'Question 784', the hole-through-the-Earth problem (Fig. 137).

It was set under the pseudonym 'Terricola' by the British mathematician Nevil Maskelyne, who was Astronomer Royal at the time.

The problem was eventually solved by Mr Robert Phillips of St Agnes in Cornwall, and his solution was duly published in *The Ladies' Diary* the following year. He found, in particular, the formula for the journey time to the Earth's

The LADIES Diary:
OR, THE
Woman's ALMANACK,
For the YEAR of our LORD, 1740.
Being the BISSEXTILE, or LEAP-YEAR:
Containing many Delightful and Entertaining *Particulars*, Peculiarly
Adapted for the *Use* and *Diversion* of the
FAIR-SEX.
Being the *Thirty-Seventh* ALMANACK ever Publish'd of this Kind.

1. HAIL! happy LADIES of the *BRITISH* Isle,
On whom the GRACES and the MUSES smile,

2. LONG had your lovely *Shape*, and matchless *Mein*,
The Wonder of the Neighb'ring Nations been;

3. NATURE to make your *Triumph* morcompleat,
To peerless CHARMS has added piercing WIT.

4. NO more let *SCTTHIA* vaunt her FEMALE-HOST,
Nor their SEMIRAMIS th' *Assyrians* boast:
WIT join'd to BEAUTY, *Fame* shall now record;
Which lead more Captive than the Conqu'ring Sword.

Printed by *A. Wilde*, for the Company of STATIONERS, 1740.

(a)

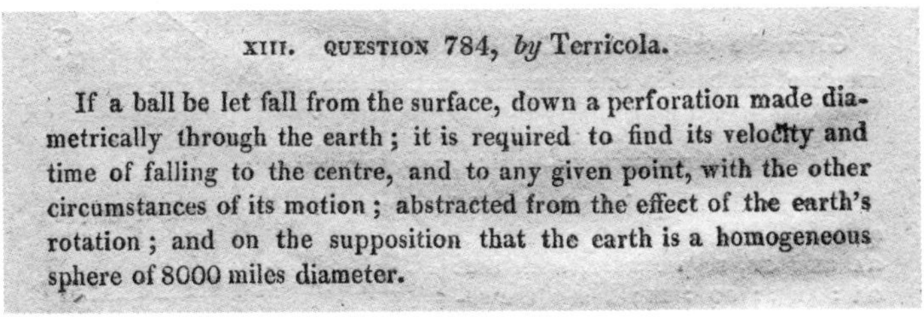

XIII. QUESTION 784, *by* Terricola.

If a ball be let fall from the surface, down a perforation made diametrically through the earth; it is required to find its velocity and time of falling to the centre, and to any given point, with the other circumstances of its motion; abstracted from the effect of the earth's rotation; and on the supposition that the earth is a homogeneous sphere of 8000 miles diameter.

(b)

Fig. 137 (a) *The Ladies' Diary.* (b) Problem 784.

centre, which is just half the transit time noted above, i.e. 21 minutes.

Yet it was not until 1882 that the French engineer Édouard Collignon showed that the transit time of just over 42 minutes is the same for *all* straight tunnels through the Earth (Fig. 138).

Ce temps n'est pas bien long: il est donné par la formule

$$T = \pi \sqrt{\frac{R}{g}};$$

si l'on y fait R = 6,366,400 mètres, rayon moyen du globe terrestre, et g = 9, 8088, on trouve pour T la valeur

42 minutes 11 secondes,

ce qui est une durée très réduite pour des voyages comme celui de Londres à Melbourne, par exemple, ou de Paris à Rio Janeiro.

Fig. 138 The transit time, from Collignon's paper of 1882.

And to understand how he did this, we need first to review some aspects of universal gravitation that had been discovered, much earlier, by Isaac Newton . . .

* * * * * * *

Newton's key idea, expressed in modern terms, is that any two point masses m and M attract one another with a force

$$F = \frac{GMm}{r^2},$$

where r is the distance between them and G is the 'universal constant of gravitation', with a value of 6.7×10^{-11} in SI units.

And he showed that this result implies that a uniform sphere of mass M and radius R exerts an external force just as if its whole mass were concentrated at the centre (Fig. 139).

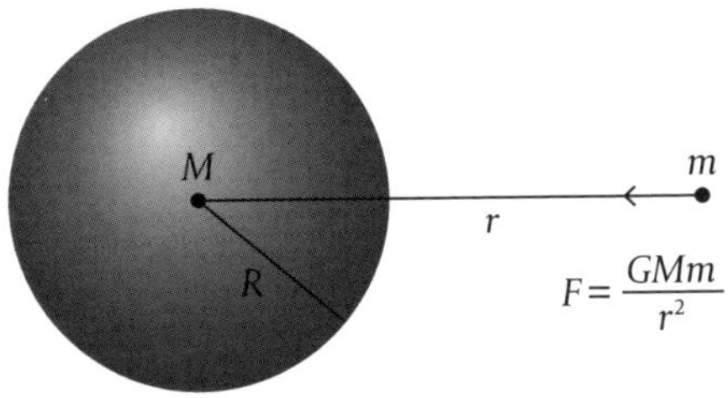

Fig. 139 The external gravitational force due to a uniform sphere.

So, when we write the force on an object of mass m close to the surface of the Earth as $F = mg$, we have

$$g = \frac{GM}{R^2},$$

where $M = 6.0 \times 10^{24}$ kg is the mass of the Earth and $R = 6.4 \times 10^6$ m its radius.

And this is how g works out to be 9.81 ms^{-2}.

* * * * * * *

Yet things proceed very differently as soon as we venture *inside* this rather simplified Earth, for then some parts of the Earth pull us upward, rather than downward, and, as a result, the force due to gravity turns out to be proportional to the distance r from the centre. We may therefore write the gravitational force on an object of mass m *inside* the Earth as

$$F = mg\frac{r}{R},$$

where R is the Earth's radius, and g the acceleration due to gravity *at its surface*.

And at this point we are ready to consider the dynamics of free fall down a tunnel.

* * * * * * *

First, let y denote the distance of the object P from the mid-point of the tunnel, as in Fig. 140. Then, as the gravitational force on the object acts towards the centre O, its component along the tunnel will be $F \cos \theta$. But $r \cos \theta = y$, so this component of force can be written instead as mgy/R.

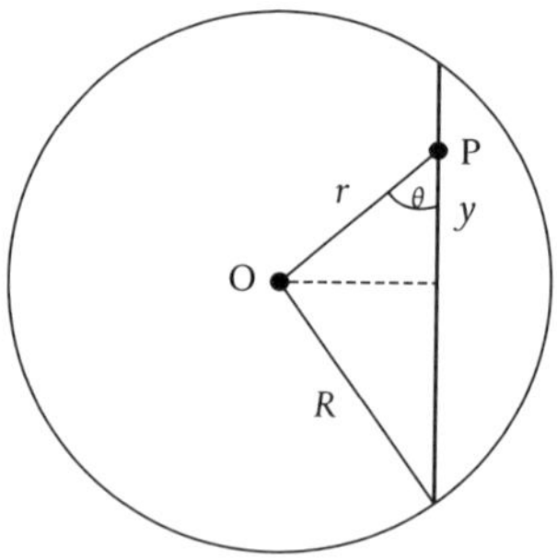

Fig. 140 Tunnel geometry inside the Earth.

When we apply the law *force* = *mass* × *acceleration*, we therefore obtain the differential equation

$$\frac{d^2y}{dt^2} = -\frac{g}{R}y.$$

And, most remarkably, this is essentially *the same differential equation that we met for the toy spider in Chapter 12*, but with $\omega^2 = g/R$.

In consequence, we know from the results there that if we start at rest at one end of the tunnel, we will oscillate between the two ends, and that the time for each complete oscillation will be $2\pi\sqrt{R/g}$, implying that the one-way transit time is $\pi\sqrt{R/g}$.

And the reason why Oxford to London would take just as long as Oxford to Sydney is that with a very short tunnel the component of gravity *along the tunnel* is also very small.

In this particular respect, in fact, there are even distinct echoes here of Galileo's chord theorem of Chapter 3, where the invariance of the travel time came about for a very similar reason.

35

TREMORS IN THE NIGHT

On the night of 7 June 1931 the UK had its biggest earthquake since records began.

The epicentre was at the Dogger Bank, about 60 miles off the Yorkshire coast in the North Sea. But its effects were even documented by the geophysicist Harold Jeffreys, in bed at Cambridge:

> I was lying approximately northeast to southwest . . . and the motion began with rapid oscillations along the bed.

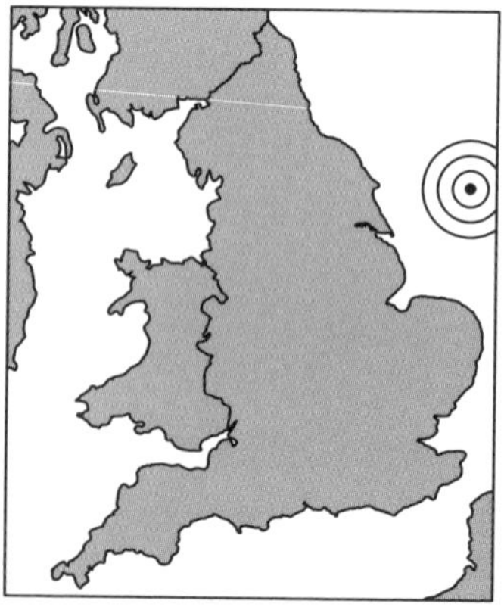

Fig. 141 The Dogger Bank earthquake of 1931.

After some time it changed to a larger but slower motion across the bed.

And, as Jeffreys observed at the time, this was all in accord with the theory of elasticity.

To see why, however, we need to go back a bit . . .

Linear elasticity

When an elastic solid is only slightly deformed, the internal stresses are typically proportional to the size of the deformation, as in Hooke's famous law.

This is so-called *linear elasticity*, and early in the nineteenth century the whole theory acquired a new lease of life with the arrival of the general differential equations, largely due to Claude-Louis Navier, Augustin-Louis Cauchy, Siméon Poisson, and—finally—George Stokes, in 1845. And two key parameters emerged from the theory, both with dimensions of stress (i.e. force per unit area).

The first is the so-called *shear modulus* μ, which is a measure of the resistance of the material to any kind of deformation.

In Fig. 142, for instance, we see a material being sheared in such a way that there is a displacement to the right by an amount q which depends only on the coordinate y. Each layer then exerts an elastic stress $\mu dq/dy$ on the material below.

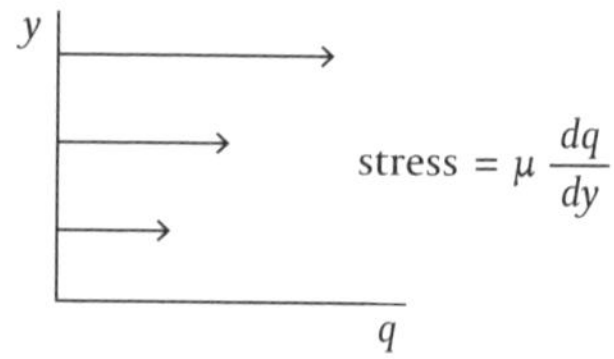

Fig. 142 Shear stress in an elastic solid.

And as this is reminiscent of the way that viscous stresses arise in fluid mechanics (Fig. 94), it is perhaps appropriate that the same symbol μ is used.

The second parameter is the *bulk modulus* λ, which comes into play only if the deformation involves changes in volume. This happens, for instance, if we take a cylinder of circular cross-section and stretch it, by applying forces to its ends. The fractional increase in length divided by the fractional decrease in radius then turns out to be a constant for the particular material in question, known as

$$\textit{Poisson's ratio} = \frac{\lambda}{2\,(\lambda + \mu)}.$$

And this time λ does play a role, because there is a change in volume associated with the deformation.

Typically, the two moduli μ and λ are of similar order of magnitude, so that Poisson's ratio is significantly less than 0.5. But there are some materials, such as India rubber, which can be sheared relatively easily but which strongly resist changes in volume, and in those cases λ is much larger than μ, so that Poisson's ratio is very nearly 0.5.

And all these ideas enter *dynamics* in a roughly similar way . . .

Elastic waves

In 1830 Poisson investigated the propagation of *waves* through elastic materials, and identified two quite different types, though the form in which we know them today is due to Stokes, 15 years later.

The first type is a shear (or 'S') wave, with motions at right angles to the direction of travel and therefore involving no changes in volume. This travels at the

$$\text{'S' wave speed} = \sqrt{\mu/\rho},$$

where ρ denotes the material density (mass/volume).

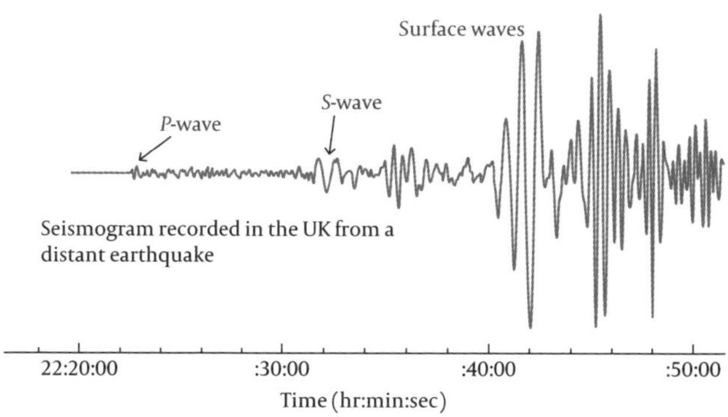

Fig. 143 A typical seismograph, courtesy of the British Geological Survey.

The second type ('P' wave) involves motions in the direction of travel, and therefore compression and expansion of the material, a bit like sound waves. This wave travels faster:

$$\text{'P' wave speed} = \sqrt{(\lambda + 2\mu)\,/\,\rho},$$

and neatly explains why Harold Jeffreys found his bed first shaking along its length when the Dogger Bank earthquake struck in 1931.

Exploring the Earth

The P and S waves caused by earthquakes have, in fact, proved to be invaluable tools for exploring the deep interior of the Earth (Fig. 144). In particular, a study by Richard Oldham in 1906 provided clear evidence for a major transition of some kind roughly halfway to the Earth's centre, and it was Jeffreys himself who, 20 years later, established definitively the existence of a liquid core.

And a major piece of evidence for *that* stems from the fact that *S waves cannot travel through a liquid*, for they are never detected at the Earth's surface beyond an angle of about 105° from the epicentre.

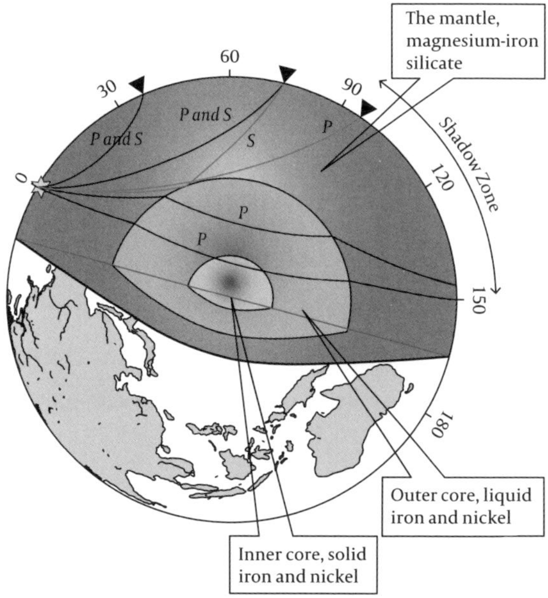

Fig. 144 Elastic waves in the Earth, and seismic evidence for the Earth's liquid core, courtesy of the British Geological Survey.

36

VERY VISCOUS FLOW

There is a famous experiment, dating from the 1960s, in which a very viscous fluid is confined in the gap between two cylinders (Fig. 145).

A blob of fluid is then marked with dye, and the inner cylinder is turned, slowly, through four complete revolutions, so that the resulting shearing motion stretches the blob into what appears to be a thoroughly mixed and faintly dyed ring. The inner cylinder is then turned *back* through four complete revolutions, and the dyed portion returns, very nearly, *to its original state as a concentrated blob.*

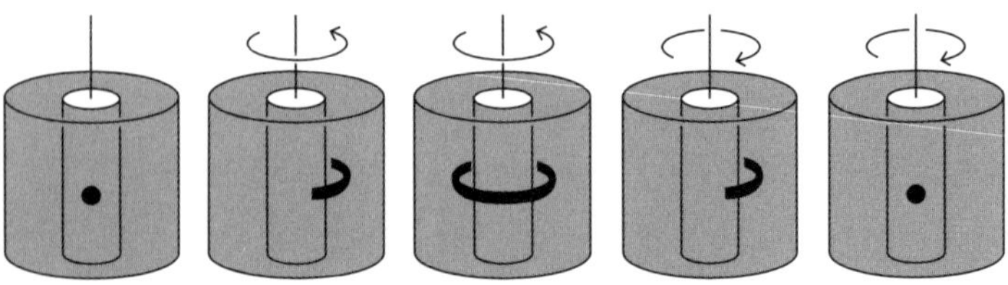

Fig. 145 Reversibility in very viscous flow.

* * * * * * *

In general, viscous fluid flows are certainly *not* reversible. This is essentially because when we change the sign of the flow everywhere, viscous terms like $\mu\partial^2 u/\partial x^2$ change sign in the equations of motion, but nonlinear terms like $\rho u \partial u/\partial x$ do not, so that the reversed flow is no longer a solution.

But if the viscosity μ is so large that the nonlinear terms are negligible anyway, the problem does not arise, and the reversed flow, in conjunction with a reversed pressure gradient, then satisfies the equations. And this is essentially what is happening in Fig. 145.

As we saw in Chapter 25, however, it is not, in fact, the viscosity itself that really matters, but rather the Reynolds number

$$R = \frac{\rho U L}{\mu},$$

and it is when this is much less than 1 that viscous forces typically become completely dominant.

* * * * * * *

An altogether different way in which R can be very small, leading to reversibility, is therefore with flow *on a tiny length scale L*.

A spermatozoon, for instance, with a typical length scale of 0.05 mm and swimming speed of 0.2 mm/s, gives rise to a Reynolds number of order 0.01. And the reversibility of flows at low Reynolds number presents problems, in fact, for the propulsion at such microscopic scales.

For a fish can swim, in principle, by simply flapping a rigid tail to and fro. But if a microscopic organism were to try the same thing it would never get anywhere, because whatever it achieved with one flap would be immediately undone by the reverse flap that followed.

Fig. 146 How spermatozoa swim.

And spermatozoa get round this problem by sending helical waves down their tails instead, as in Fig. 146b.

* * * * * * *

There is, in addition, one rather more obscure way in which viscous forces can dominate a flow, and this appeared in an article by the English engineer Henry Hele-Shaw for the magazine *Nature* in 1898.

The fluid in question was simply water, flowing with uniform speed U past an object of uniform cross-section and horizontal length scale L, but the novelty here is that the whole flow takes place in the *very narrow gap, height h, between two glass plates* (Fig. 147a).

This creates such strong shear between the plates that it is necessary only for the so-called 'reduced Reynolds number'

$$\frac{\rho UL}{\mu}\left(\frac{h}{L}\right)^2$$

to be small for viscous forces to completely dominate, even with a fluid of low viscosity, like water. And this can evidently always be achieved, in principle, simply by making h/L small enough.

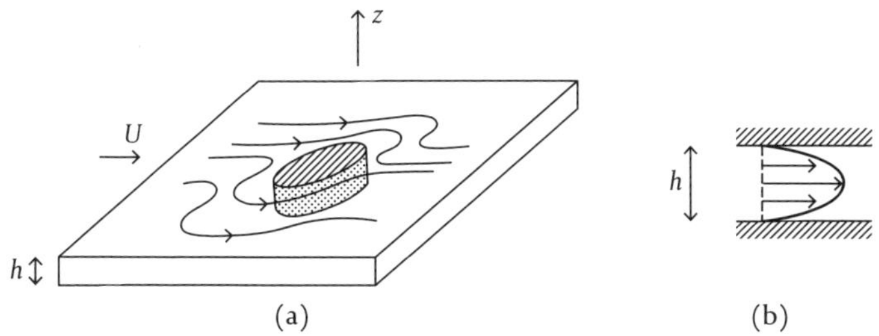

Fig. 147 Flow in a Hele-Shaw cell.

The flow speed then varies with height everywhere in the way indicated in Fig. 147b, satisfying the no-slip conditions on the top and bottom plates.

Yet while the flow *speed* varies with height, it turns out, quite remarkably, that the streamline pattern does not. And, more remarkably still, when we look down on the apparatus from above the flow pattern that we see past the object in question is *the one predicted by nonviscous theory!*

In particular, the theoretical nonviscous flow past a flat and steeply inclined 'wing', in Fig. 148a, which would normally be completely unobservable because of viscous boundary layer separation at the sharp edges, is exactly what one sees in this strange apparatus, and Fig. 148b is, in fact, Hele-Shaw's original photograph of the phenomenon.

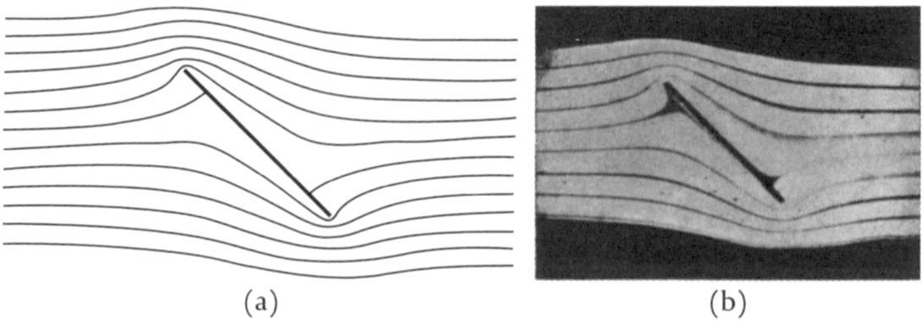

Fig. 148 (a) Theoretical nonviscous flow, without circulation, past a flat 'wing'. (b) The equivalent flow in a Hele-Shaw cell.

And there has to be a certain irony, I think, when the only way we can actually observe some of the more outlandish predictions of nonviscous flow theory is in an experimental arrangement which is completely dominated by the effects of viscosity!

37

WHICH WAY DOES THE BATHWATER SPIN?

In the early 1960s a team at MIT in the United States sought the definitive answer to an old, old question: which way does the bathwater spin when you pull the plug out?

And, in particular, does it depend on which hemisphere you're in, because of the Earth's rotation?

Fig. 149 The bathtub vortex.

They set up a circular tank, 6 feet in diameter and 6 inches high, with a small drain hole 3/8 inch in diameter. And to minimize disturbances caused by pulling the plug they attached to the drain hole a 20-foot rubber hose, with the plug at the far end.

They then filled the tank, quite deliberately, with the water swirling *clockwise*, i.e. in the opposite direction to any motion that might be induced by the rotation of the Earth. Finally, they covered the tank with a plastic sheet to eliminate any effect of air currents, and let the water settle for 24 hours.

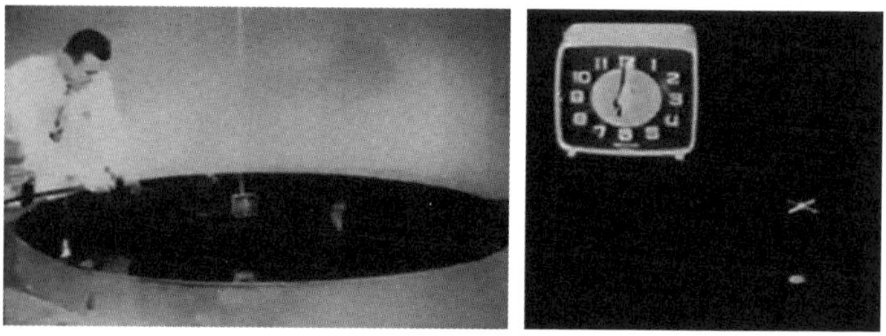

(a) (b)

Fig. 150 The MIT bathtub experiment. In (b) the vorticity float can just be seen, directly above the drain hole.

The next day, they gently lowered a tiny 'vorticity float', consisting of two small vanes at right angles, onto the water surface, directly above the drain hole, to visualize any spin of the fluid. And, finally, they pulled the plug from the end of the long hosepipe.

For at least 10 minutes, nothing happened to the vorticity float at all—it remained stationary.

Then, after about 15 minutes the experimenters detected a tiny *anticlockwise* rotation. And a few minutes later still, when the tank had almost drained, the vorticity float was rotating anticlockwise once every 3 or 4 seconds.

These, then, are the kind of lengths you have to go to if you want the Earth's rotation to determine which way the bathwater spins, and this was confirmed three years later when another team repeated the experiment in Sydney, Australia, and found that the same routine led to a spin in the opposite direction.

In normal, everyday circumstances, on the other hand, any effect of the Earth's rotation on the bathtub vortex is completely dwarfed by other influences, and it is not difficult to see why, from the mathematics.

When is the Coriolis force significant?

We noted in Chapter 30 that the *Coriolis force* per unit mass due to the Earth's rotation is $2\Omega v \sin\phi$ at latitude ϕ, and tends to deflect a fluid motion to the right in the northern hemisphere but to the left in the southern hemisphere, where ϕ is negative.

So, if V denotes a typical flow speed relative to the Earth and L a typical length scale of the flow, the Coriolis force per unit mass will be of order $2\Omega V$, while—as we have seen before—a typical acceleration relative to the Earth will be of order V^2/L, so that

$$\frac{\textit{Coriolis force per unit mass}}{\textit{fluid acceleration}} \approx \frac{2\Omega L}{V}.$$

Now, Ω, the angular velocity of the Earth, is approximately 7.3×10^{-5} radians/sec. So, if we take 5 cm per second as a typical flow speed and 30 cm as a reasonable length scale for the bathtub vortex, $2\Omega L/V$ is about 0.001, indicating that Coriolis effects due to the Earth's rotation will usually be negligible and dwarfed by other influences on the fluid motion.

The Taylor–Proudman theorem

It is perfectly possible, however, to make Coriolis effects completely dominant in fluid mechanics on a laboratory scale, simply by putting a tank of water on a rapidly rotating turntable, and then creating some *slow* fluid motion relative to the tank itself.

Moreover, something quite extraordinary then happens.

Suppose, for instance, that we arrange for a small object to be towed slowly and steadily along the bottom of the tank. Needless to say, if the tank were stationary the disturbance to the fluid would be fairly localized, with hardly any fluid motion at all near the top of the tank.

But suppose now that the whole apparatus is rapidly rotating, and that the towing speed V of the object (relative to the turntable) is much smaller than a typical speed of rotation, so that the ratio $\Omega L/V$ is much greater than 1.

Then the whole column of fluid above the object *moves with it across the tank* (Fig. 151a), and the rest of the fluid has to go round that column, just as if the object itself extended upward to the top of the tank.

Fig. 151 A Taylor column in a rapidly rotating fluid caused by (a) a slowly moving object and (b) a slight spin-down of the container.

This is a consequence of a result known as the Taylor–Proudman theorem (Notes, p. 253), and G. I. Taylor himself wrote at the time

> This idea appears fantastic

before going on to explain that the experiments confirm that it really does happen.

Geophysical and astrophysical fluid mechanics

Coriolis effects due to the Earth's rotation are usually negligible for the bathtub vortex essentially because the length scale L of the motion is far too small, but for fluid motions on a planetary scale it's an altogether different story.

Why is it, for instance, that in TV weather forecasts the airflow is always roughly *along* isobars, i.e. along lines of constant pressure?

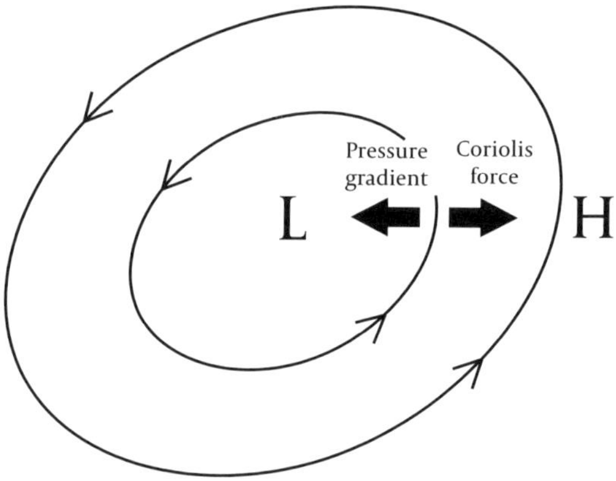

Fig. 152 The Coriolis force in meteorology.

The answer is that for large-scale atmospheric motions the parameter $2\Omega L/V$ is significantly larger than 1, so that the Coriolis force is dominant and the only force that can more or less balance the pressure gradient (Fig. 152). And as the Coriolis force acts at right angles to the flow velocity itself, that flow must therefore be *along* the lines of constant pressure.

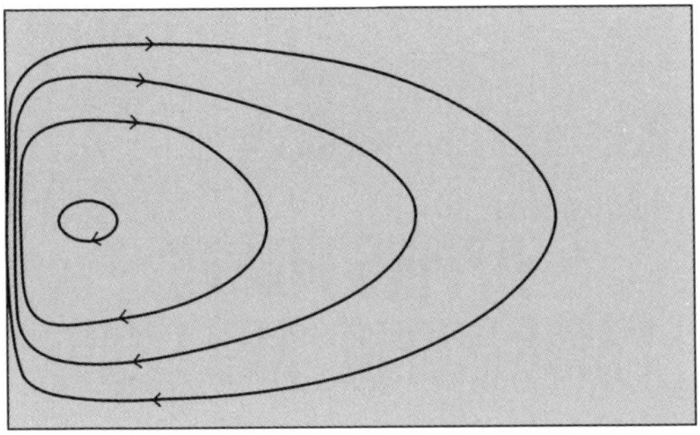

Fig. 153 A simple model of wind-driven ocean circulation in the northern hemisphere, displaying western intensification.

Moreover, as the Coriolis force acts to turn the flow to the right in the northern hemisphere, it explains why the flow around an area of low pressure there is anticlockwise.

In the oceans, too, the Coriolis force can play a major role. In particular, the way that intense currents like the Gulf Stream form on *western* boundaries in the norther hemisphere is down to a very subtle effect which arises from the increase in Coriolis force $2\Omega v \sin\phi$ with latitude (Fig. 153).

And on an astrophysical length scale, such as one finds in the internal dynamics of the Sun and the stars, where the effects of magnetic fields and electric currents on fluid motion often have to be taken into account, the Coriolis force can frequently play an even more dominant role.

38

TIPPING POINTS, CATASTROPHE, AND CHAOS

In recent years, two big dynamical ideas seem to have struck a chord with the general public.

The first is the idea of a *tipping point*, especially in relation to the atmosphere, oceans, and climate change. This is where a very small change in the operating conditions of a system can cause it to suddenly *jump* to some completely different state.

A simple example is shown in Fig. 154, where an elastic strut bends under a heavy weight.

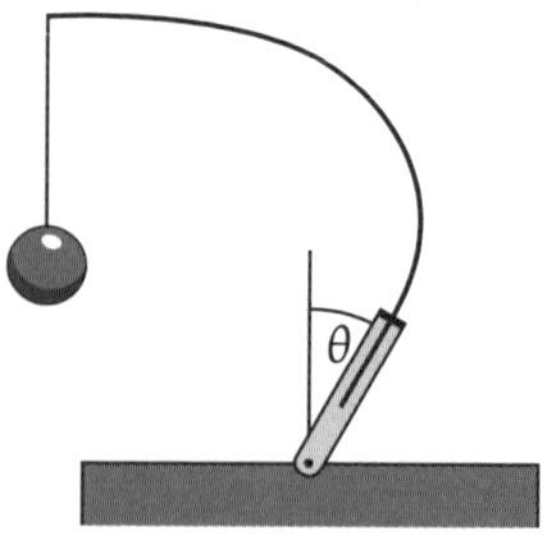

Fig. 154 A tipping point in elasticity.

At first, as we gradually increase the clamp angle θ, the shape of the strut changes smoothly and gradually. But when the clamp angle exceeds some critical value the strut suddenly 'jumps' to a completely different state, bending deeply to the right instead.

Moreover, as soon as this has happened, the system does *not* immediately jump back if we start decreasing θ. There *is* eventually a jump back, but only when θ has been gradually reduced below a second (and negative) critical value.

This behaviour is an example of *hysteresis*, and only happens because there is a range of θ for which the system has *two* stable steady states—one leaning to the left and one leaning to the right—separated by an unstable one. And this, in turn, only happens if the weight is big enough to cause buckling instability of the vertically upright state when $\theta = 0$, in the manner we saw in Fig. 86. If the weight

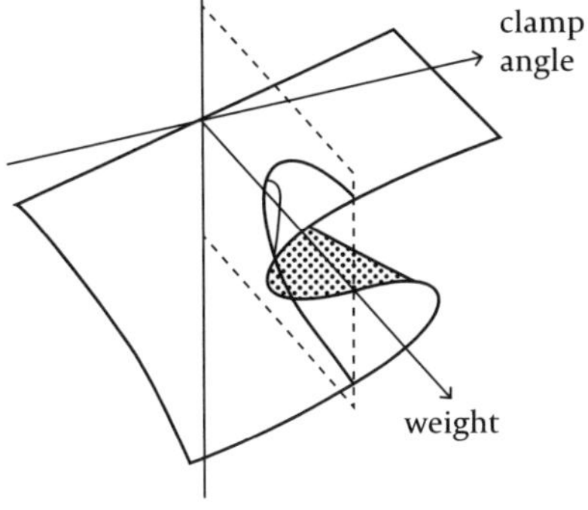

Fig. 155 Catastrophe theory at work in elasticity.

is below that critical value, there is only one static state for each value of θ, and the whole 'jump' phenomenon disappears.

And all this is summarized schematically in the response diagram of Fig. 155, which is one of the most iconic figures in *catastrophe theory*.

* * * * * * *

The second major idea to grab public attention is *chaos*, where a completely deterministic system oscillates, but in a seemingly haphazard manner.

These ideas emerged most famously in a 1963 paper by the American meteorologist Ed Lorenz.

He was studying a highly simplified mathematical model of thermal convection in a layer of fluid heated from below, and the computer solutions of the differential equations revealed oscillations with time which never settled down to a regular, repeating pattern.

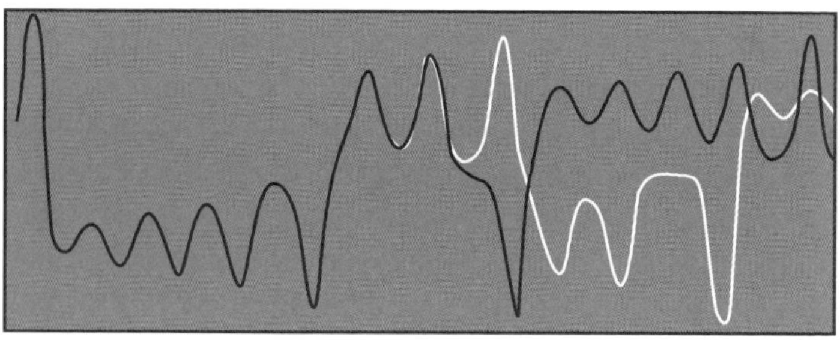

Fig. 156 Chaos and sensitivity to initial conditions in the Lorenz equations.

Moreover, the way in which the irregular oscillations evolved with time depended very sensitively on the initial conditions. In Fig. 156, for example, we see one of the variables in the Lorenz equations plotted against time for two different initial conditions. And while those initial conditions differ by only 1 part in 1000, after a small number of oscillations the system plainly evolves in two entirely different ways.

Lorenz was well aware of the possible implications for weather forecasting, writing

> When our results . . . are applied to the atmosphere . . . they indicate that prediction of the sufficiently distant future is impossible by any method, unless the present conditions are known exactly. In view of the inevitable inaccuracy and incompleteness of weather observations, precise very-long-range forecasting would seem to be non-existent.

Yet the full significance of his work was not immediately appreciated at the time, partly because his particular differential equations were known to represent a rather crude model of atmospheric dynamics.

What scarcely anybody seemed to realize was that Lorenz's findings would prove to be fairly *typical* of a wide variety of nonlinear dynamical systems with a certain minimal degree of complexity.

* * * * * * *

There had, in fact, been one or two earlier intimations of chaos—and not only with Poincaré's classic work in 1889 on the gravitational three-body problem.

In 1958, for instance, D. W. Allan, then at Cambridge but later at King's College London, published a short paper in *Nature* about reversals of the Earth's magnetic field, which is thought to be maintained by motions in the liquid core (see Fig. 144). Each reversal takes place on a time scale of a few thousand years, but the reversals themselves occur only erratically, with much longer periods between each one where the magnetic field remains of one sign.

At the time of Allan's paper, the fluid mechanics of the whole problem was totally intractable, even with the help of electronic computers, and much simpler 'disk dynamo' models, in which the fluid core is replaced by a rotating conducting disk, were of considerable interest.

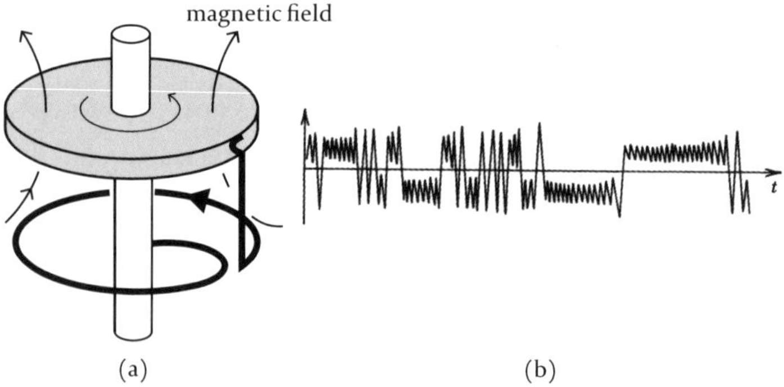

Fig. 157 (a) The classic disk dynamo. (b) Current reversals in the model of Robbins (1977), where the classic model is modified by the addition of a shunt.

In the classic model of Fig. 157a, the disk is driven by a torque, and its motion in the magnetic field creates an electric current which is fed into a coil so as to produce a magnetic field which reinforces the original.

Allan showed that when two such disk dynamos are coupled together the magnetic fields would occasionally reverse in sign, and in an irregular, nonperiodic way.

Moreover, a later study by Kay Robbins, then at MIT, showed that a single disk dynamo could show similar behaviour, when suitably modified. And her computer solutions of the differential equations show clearly the comparatively long periods of constant polarity between each reversal, and the erratic way in which the reversals themselves occur (Fig. 157b).

* * * * * * *

Yet you don't need anything as exotic as this to see chaos in action.

You may recall, for instance, that in 1738 Daniel Bernoulli studied the small oscillations of a double pendulum, and found some very neat periodic solutions.

Yet with a *large* initial disturbance everything changes dramatically, and the whole system provides a most evocative demonstration of chaos.

Figure 158b shows a particularly well-engineered tabletop example, in which the outer pendulum mass is much the smaller of the two. And if I release the pendulums from an aligned and horizontal position, the outer one

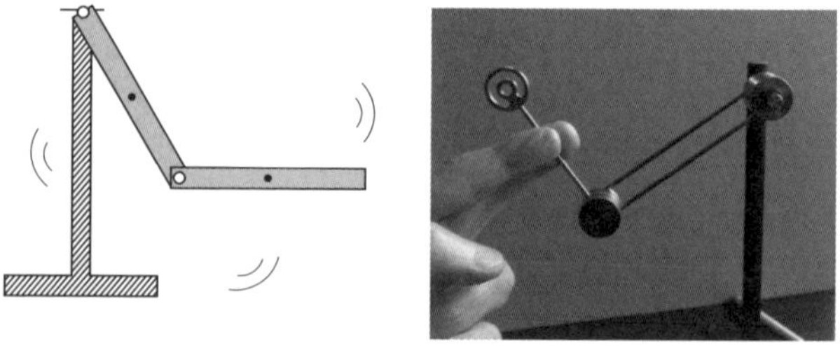

(a) (b)

Fig. 158 Large-amplitude motion of a double pendulum.

gets thrown about in a haphazard sequence of whirling 'episodes'. In each individual episode the whirling motion is either clockwise or anticlockwise, with the number of rotations varying greatly from one episode to another. And there are typically a dozen or so such episodes before friction eventually takes its toll and the pendulums end up hanging downward.

Moreover, the extreme *sensitivity to initial conditions*, so characteristic of chaos, is also very evident in this simple apparatus. For I actually possess *two* identical systems of this type, and if I set them both going, from initial conditions that appear almost identical, they stay 'in harmony' for only a second or two before the whole sequence of episodes evolves completely differently in the two cases.

39

AND FINALLY . . .

We have seen many times in this book how an old problem in mechanics can often acquire a new and unexpected 'twist', long after it first came to life, and I would like to end with a further short selection.

The gravitational *three-body problem*, for instance, played a fundamental role in the Apollo Moon missions of the 1960s and '70s, and Fig. 159 shows a figure-of-eight trajectory. In this particular case the spacecraft arrives just ahead of the Moon in its orbit, and exploits a 'slingshot' effect due to the Moon's gravity to get propelled back to Earth.

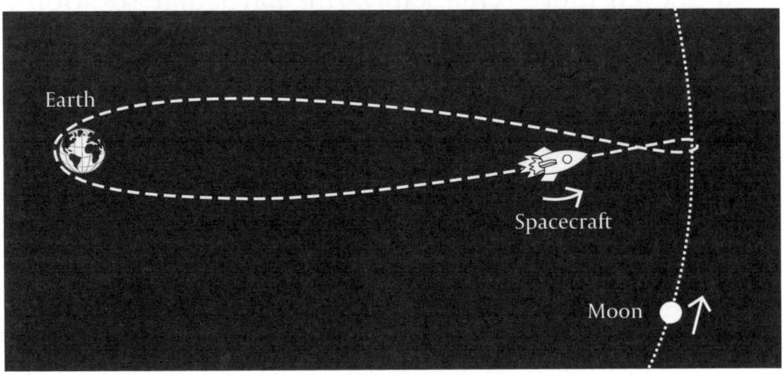

Fig. 159 A typical 'free return' path for an Apollo spacecraft.

Considerations of fuel made 'free return' trajectories like this especially important in the space programme, and during the ill-fated Apollo 13 mission they played a crucial part in getting the astronauts safely home.

As the mass of the spacecraft is tiny in comparison with that of the Earth or the Moon, the 'restricted' version of the three-body problem applies, but the full problem has also seen some new twists and turns in recent years.

In 1993, for instance, Cris Moore of the Santa Fe Institute discovered a new solution in which three equal gravitating masses *chase each other round a figure of eight* (Fig. 160).

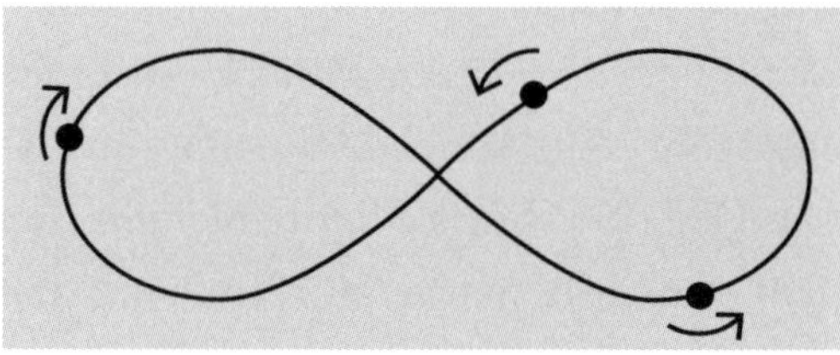

Fig. 160 Figure-of-eight motion in the three-body problem.

And, even more remarkably, this particular solution is stable, in the sense that if one of the masses starts slightly out of place, the figure of eight doesn't break up—it just slowly *rotates*, maintaining its overall form.

* * * * * * *

Research in fluid mechanics, too, is as lively as ever.

I hope, for instance, that I did not give any impression earlier that highly viscous flows are always fairly 'simple', because as soon as a free boundary is involved (which must be determined as part of the problem), all hell can break loose.

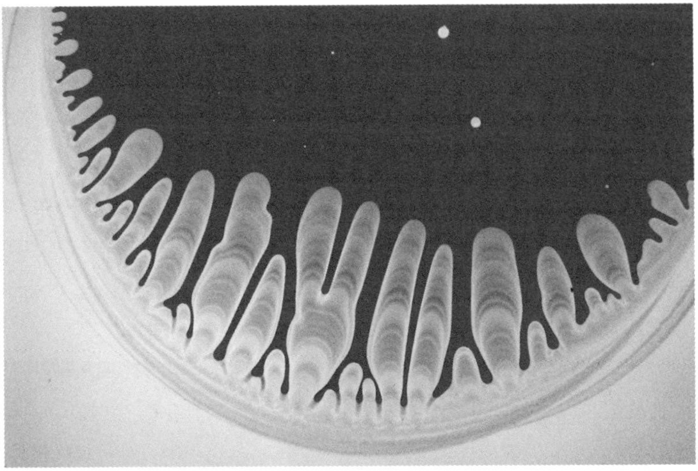

Fig. 161 An instability of very viscous flow.

Take, for instance, two sheets of glass or Perspex, and put a blob of washing-up liquid on one of them. Then press down with the other sheet, so that the gap between the sheets becomes very narrow and viscous forces predominate. The blob will then spread out in a nice, symmetric fashion, with a more-or-less circular boundary.

Now gradually pull the sheets apart again.

The *reverse* motion will be hopelessly unstable, so that tiny ripples appear in the boundary—for no apparent reason—and then grow rapidly into long viscous 'fingers' (Fig. 161). And if you proceed further still, these fingers will, themselves, split into others in a roughly fractal-like way.

* * * * * * *

I would like to end, however, with a surprising twist to one of the oldest scientific subjects of all—the pendulum.

For on a wet Sunday afternoon in 1992 I took Daniel Bernoulli's multiple-pendulum system of 1738 and proved that the whole system can be stabilized *upside-down* by vibrating the pivot up and down by a small enough amount and at a high enough frequency (Fig. 162).

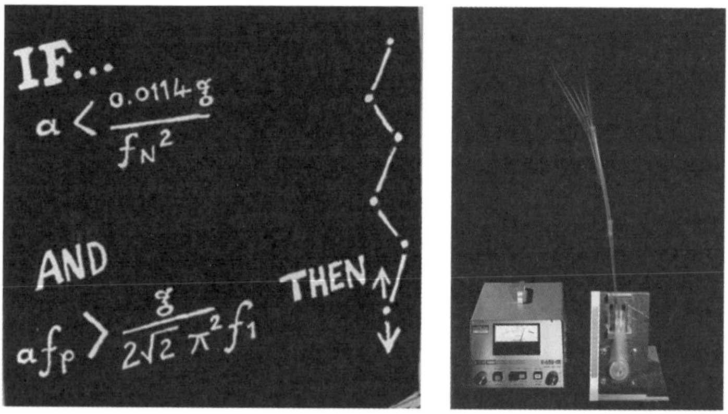

Fig. 162 (a) The upside-down pendulum theorem (1993). Here a denotes the amplitude of the pivot vibration and f_p its frequency. The two numbers f_1 and f_N denote the lowest and highest frequencies of small oscillation of the system about the downward vertical when the pivot is fixed, as investigated by Daniel Bernoulli in 1738. (b) An actual experiment, showing a stabilized upside-down triple pendulum. (See also Notes, p. 255.)

The idea that this can be done for a single rigid pendulum was, in fact, well known at the time, and dated back to a paper in 1908 by Andrew Stephenson, a lecturer at Manchester University. But the new theorem showed that the same 'trick' can also be done with *any finite number* of linked pendulums, all balanced on top of one another.

Moreover, Tom Mullin, a colleague of mine in Oxford at the time, confirmed the predictions of the theorem experimentally, and we were even a little surprised at just how robustly stable the inverted pendulums could be. Figure 162b, for instance, shows a triple-pendulum system gradually wobbling back to the upward vertical after an initial displacement of some 45°.

And while I don't delude myself that the upside-down pendulum theorem has anything like the importance of most of the other results in this book, it is, surely, yet one more example of how, even after all this time, classical mechanics is still capable of springing surprises.

Notes

1. Introduction

For biographies of Newton and Euler see

Never at Rest: A Biography of Isaac Newton, by Richard Westfall (Cambridge University Press, 1980);
Leonhard Euler: Mathematical Genius in the Enlightenment, by Ronald S. Calinger (Princeton University Press, 2016).

2. The Beginning of Dynamics

For a study of the myth (or reality?) concerning Fig. 4, see

Aristotle, Galileo and the Tower of Pisa, by Lane Cooper (Cornell University Press, 1935).

For scholarly studies of Galileo's work on dynamics see

Galileo: Pioneer Scientist, by Stillman Drake (University of Toronto Press, 1990);
Galileo Galilei and Motion, by Roberto Vergara Caffarelli (Springer, 2009).

Something of Professor Caffarelli's meticulous reconstructions of Galileo's inclined plane experiments can be seen at https://www.youtube.com/@labGalileo/featured. (See, for instance, his 2005 video 'Il laboratorio di Galileo Galilei' at, in particular, 21–28 minutes.)

* * * * * * *

For the Merton rule concerning motion with constant acceleration see

The Science of Mechanics in the Middle Ages, by Marshall Clagett (Oxford University Press, 1959).

In addition, good general accounts of the early development of dynamics can be found in

The Birth of the New Physics, by I. Bernard Cohen (Penguin Books, 1987);
A History of Mechanics, by René Dugas (Dover, 1988).

3. Force, Mass, and Acceleration

The reason that the acceleration of the ball's centre down the plane is less than $g \sin \alpha$ in Fig. 13 is as follows.

As the centre of the ball accelerates, the ball itself has to rotate faster, so that its angular momentum increases with time. By Chapter 16 this requires a torque about the ball's centre, which can only be supplied by a frictional force at the point of contact with the plane, and directed *up the plane*. This reduces the net force on the ball acting down the plane, leading to a smaller acceleration.

* * * * * * *

To understand Galileo's chord theorem, it is simplest if we consider sliding down a frictionless plane, rather than rolling. Then, with reference to Fig. 163, the time T of descent down a plane of length L can be obtained from

$$L = \frac{1}{2} g' T^2,$$

where $g' = g \cos A$ denotes the component of gravity down the plane. But the angle subtended on a circle by a diameter is a right angle (Thales' theorem), so $\cos A = L/D$, where D is the diameter. The factors of L then cancel, and $T = \sqrt{2D/g}$, the same for all such chords.

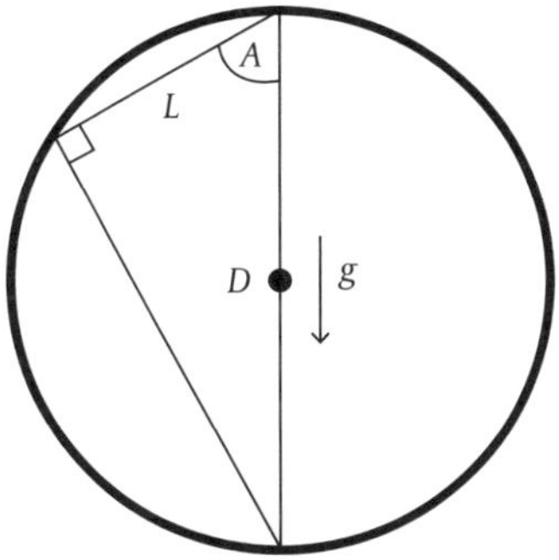

Fig. 163 Illustrating Galileo's chord theorem.

4. Circular Motion

For further discussion of Newton's argument in Fig. 18, see

The Background to Newton's Principia, by John Herivel (Clarendon Press, 1965), especially p. 193.

As angle OAB in Fig. 18 is a right angle, a simple alternative way of calculating h is to use Pythagoras's theorem:

$$(h+r)^2 = r^2 + (vt)^2.$$

5. Energy

An individual who is upside-down at the top of the 'Grand Centrifugal Railway' of 1842 will need to be moving with a speed v_T there such that $\frac{v_T^2}{r} > g$ to avoid falling out at that point, in order that—from *their* dynamical point of view—the centrifugal force upward is greater than the downward force due to gravity (see Chapter 30). This is why the acceleration at the bottom will have to be at least $5g$.

6. Collisions

Galileo's passage about dynamics on a stationary or *uniformly* moving ship can be found in his

Dialogue Concerning the Two Chief World Systems—Ptolemaic and Copernican, translated by Stillman Drake (University of California Press, 1967), pp. 186–7.

Huygens actually makes extensive use of such ideas in the way he *obtains* his rules of collision, as explained in

Absolute or Relative Motion? Vol. 1: *The Discovery of Dynamics*, by Julian Barbour (Cambridge University Press, 1989), pp. 462–73.

7. 'A Scab of Symbols'

For a good history of the use of algebraic formulae in physics, and indeed of the method of dimensions, see

The Mathematics of Measurement, by John Roche (Athlone Press, 1998).

8. Reaching for the Heavens

For a good general history of mathematical astronomy see

From Eudoxus to Einstein, by C. M. Linton (Cambridge University Press, 2004).

To see how the formula v^2/r arises (in word form) in Huygens' book of 1673, see

Christiaan Huygens' The Pendulum Clock, translated by Richard Blackwell (Iowa State Press, 1986), pp. 176–8.

9. *Principia*

There are, of course, many scholarly books on different aspects of Newton's *Principia*, but for an approachable *overall* look, I suggest, as a starting point,

Magnificent Principia, by Colin Pask (Prometheus Books, 2019).

10. Newton and Planetary Motion

For more details of Newton's highly geometrical arguments relating to Fig. 40 see

Reading the Principia, by Niccolo Guicciardini (Cambridge University Press, 1999), pp. 48–56.

11. Calculus

A general introduction to calculus and its history can be found in my book *The Calculus Story* (Oxford University Press, 2017), but I also recommend

Infinite Powers, by Steven Strogatz (Atlantic Books, 2019),
Calculus Gems, by George F. Simmons (McGraw-Hill, 1992).

For an informal look at calculus applied to more general physical problems see

17 Equations that Changed the World, by Ian Stewart (Profile Books, 2012).

* * * * * * *

To calculate the downward speed in connection with Fig. 42, note first that after a small *extra* time t', the total distance fallen will be

$\frac{1}{2}g(t+t')^2$, which is equal to $\frac{1}{2}g\left(t^2+2tt'+t'^2\right)$. The small *extra* distance fallen is therefore $\frac{1}{2}g\left(2tt'+t'^2\right)$, and dividing by the small extra time t' then gives $g\left(t+\frac{1}{2}t'\right)$. In the final step, when we choose t' to be smaller and smaller, this gets closer and closer to gt, which must therefore be the downward speed actually at time t.

12. Oscillations

If $y = a\cos\omega t$, then, from elementary calculus, $\frac{dy}{dt} = -a\omega\sin\omega t$, and therefore

$$\frac{d^2y}{dt^2} = -a\omega^2\cos\omega t = -\omega^2 y.$$

So $y = a\cos\omega t$ satisfies the differential equation in Fig. 45 for all time t, and it also satisfies the initial conditions, because $y = a$ and $\frac{dy}{dt} = 0$ when $t = 0$.

13. A New Age Begins . . .

In Fig. 48, ddx/dt^2 is Euler's notation for what we would now write as

$$\frac{d^2x}{dt^2} = \frac{d}{dt}\left(\frac{dx}{dt}\right).$$

The strange factor of 2 arises because some of the symbols haven't quite got their modern meanings. In particular, Euler's M is actually our *weight* Mg, and he chooses, in effect, units of length and time such that the numerical value of g is $1/2$.

* * * * * * *

To see roughly why the result in Fig. 50 implies that OP sweeps out area at a constant rate, note first that after a short extra time t' the mass will have moved a short distance $v_\theta t'$ in the transverse direction. While it will also have moved a short distance in the radial direction, the extra area traced out will nevertheless be very nearly that of a thin isosceles triangle with height r and base $v_\theta t'$, and therefore area $\frac{1}{2} r v_\theta t'$. And on dividing by the short extra time t', the *rate* at which area is being swept out emerges as $\frac{1}{2} r v_\theta = \frac{1}{2} r^2 d\theta/dt$.

14. 'Contrary to Y^e^ Opinion of Y^e^ Vulgar'

Newton's argument leading to his 'slightly to the east' conclusion is that the Earth's gravity acting on the dropped object

> will give it a new motion towards ye center of ye Earth without diminishing ye old one from west to east. Whence the motion of this body from west to east, by reason that before it fell it was more distant from ye center of ye earth than the parts of ye earth at wch it arrives in its fall, will be greater than the motion from west to east of ye parts of ye earth at wch ye body arrives in its fall.

And, in my opinion—for what that's worth—this reads as if Newton is assuming, in modern terms, that, relative to a stationary frame and polar coordinates, it is the tangential velocity of the object $rd\theta/dt$ that is constant during the motion, rather than $r^2 d\theta/dt$, and this leads to a small total displacement to the east that is only half the correct answer.

15. The Principle of Linear Momentum

For a very short history of mechanics that demolishes what its author calls the 'folklorish' one, and gives proper emphasis to the major role played by Euler, especially in connection with the principle of linear momentum, see

'History of Classical Mechanics, Part 1, to 1800', by Clifford Truesdell, in *Die Naturwissenschaften*, **63** (1976): 53–62.

* * * * * * *

The detailed argument relating to Fig. 54 is as follows.

Denote by $\boldsymbol{F}_{CA}$ the pressure force exerted by the fluid in region C on the fluid in region A, and so on. Denote by $\boldsymbol{L}_A$ the momentum of the fluid in region A, and similarly for region B.

Applying the principle of linear momentum to the bodies A, B, and A + B in turn leads to

$$\boldsymbol{F}_{CA} + \boldsymbol{F}_{BA} = \frac{d}{dt}\boldsymbol{L}_A,$$

$$\boldsymbol{F}_{CB} + \boldsymbol{F}_{AB} = \frac{d}{dt}\boldsymbol{L}_B,$$

$$\boldsymbol{F}_{CA} + \boldsymbol{F}_{CB} = \frac{d}{dt}\left(\boldsymbol{L}_A + \boldsymbol{L}_B\right),$$

and on adding the first two equations we have consistency only if $\boldsymbol{F}_{BA} = -\boldsymbol{F}_{AB}$.

16. The Principle of Angular Momentum,

Strictly speaking, the torque about O in Fig. 56a is actually a *vector* quantity, defined by the vector product $\boldsymbol{r}\wedge\boldsymbol{F}$, where $\boldsymbol{F}$ is the force and $\boldsymbol{r}$ the position vector with respect to O. As a vector, the torque in Fig. 56a is therefore directed upward, out of the paper.

In a similar way, the angular momentum (or 'moment of momentum') about a point O is also a vector quantity, defined for the point mass in Fig. 57, for instance, as $\boldsymbol{r}\wedge m\boldsymbol{v}$, where $\boldsymbol{v}$ denotes the velocity.

* * * * * * *

Much of the deepest scholarship on Euler's work in mechanics is by Clifford Truesdell (1919–2000), especially in his *Essays in The History of Mechanics*, published by Springer-Verlag in 1968.

Truesdell was one of the strongest expositors of the fundamental principles of linear and angular momentum as (a) applicable to bodies of all kinds and (b) independent of one another, except in certain special cases, this independence being especially evident in the case of flexible or elastic bodies.

Moreover, in one of his most remarkable essays, 'Whence the Law of Moment of Momentum?', he tentatively attributes these views very largely to Euler himself, in a paper published in 1776. Truesdell places particular emphasis on the six equations in Fig. 164, which he interprets as the first statement of the principles of linear and angular momentum together, side by side, as *independent* and *general* laws of mechanics.

§. 29. Hac igitur ratione ſex nacti ſumus aequationes, quas hic coniunctim conſpectui exponamus

$$\text{I. } \int dM\left(\frac{ddx}{dt^2}\right) = iP \qquad \text{IV. } \int z\,dM\left(\frac{ddy}{dt^2}\right) - \int y\,dM\left(\frac{ddz}{dt^2}\right) = iS$$

$$\text{II. } \int dM\left(\frac{ddy}{dt^2}\right) = iQ \qquad \text{V. } \int x\,dM\left(\frac{ddz}{dt^2}\right) - \int z\,dM\left(\frac{ddx}{dt^2}\right) = iT$$

$$\text{III. } \int dM\left(\frac{ddz}{dt^2}\right) = iR \qquad \text{VI. } \int y\,dM\left(\frac{ddx}{dt^2}\right) - \int x\,dM\left(\frac{ddy}{dt^2}\right) = iU.$$

Fig. 164 From Euler's paper 'Nova methodus motum corporum rigidorum determinandi', *Novi. Comm. Acad. Sci, Petrop.*, **20** (1776): 208–38.

He does, however, frame this interpretation as a 'conjecture', and while Truesdell's view has been enormously influential, it has also come under recent scrutiny, most notably in a major new study on Euler's developing thoughts on the whole matter:

Leonhard Euler's Principle of Angular Momentum, by Andreas Verdun (Springer-Verlag, 2025).

* * * * * * *

Today, we may obtain the differential equations of motion for a rigid body easily by applying the principle of angular momentum, which obviates all need to consider or speculate about *internal* forces between different parts of the body itself.

The 'spurious' argument that I refer to at the end of Chapter 16 concerns attempts in many textbooks of mechanics to *deduce* the principle of angular momentum for a rigid body from Newton's second law. For such attempts inevitably require some consideration of the internal forces, and typically treat the rigid body, for just this part of the argument, as a system of point masses, with equal and opposite forces between each two mass points *acting along the line joining them*.

Truesdell traces this somewhat bizarre line of reasoning back at least as far as Thomson and Tait's *Treatise on Natural Philosophy* (1867), and a century later it still featured in all my various mechanics textbooks as an undergraduate in the 1960s. I even remember feeling strangely uneasy about it at the time.

To put it bluntly, having made the uncompromising assumption of complete rigidity, no matter how the body actually moves, have we any business making assumptions about the internal forces that might be needed to achieve that?

And while I had no idea of this at the time, some rather more distinguished scholars of mechanics had already expressed similar doubts. As early as 1914, for instance, the treatment of rigid body motion in Horace Lamb's *Dynamics* includes the following:

> In the author's view it is best to postulate the principles of linear and angular momentum as such, regarding them as natural extensions of the Newtonian Laws of Motion . . . Since some assumption has in any case to be made, it seems best to make it directly in the form which is most convenient for further developments, and is at the same time independent of doubtful hypotheses.

17. Spinning Tops

For a delightfully illustrated and old-fashioned account of spinning tops, see

Spinning Tops and Gyroscopic Motion, by Harold Crabtree (Longman Green and Co., 1909).

As mentioned in Chapter 16 and the associated 'Notes', p. 236, the torque and angular momentum about a point O are actually vectors. For a rapidly spinning symmetric top, such as that in Fig. 62, the angular momentum vector $\boldsymbol{L}$ is very nearly in the direction of the top's axis, and this permits the fairly simple explanation of precession in Fig. 165.

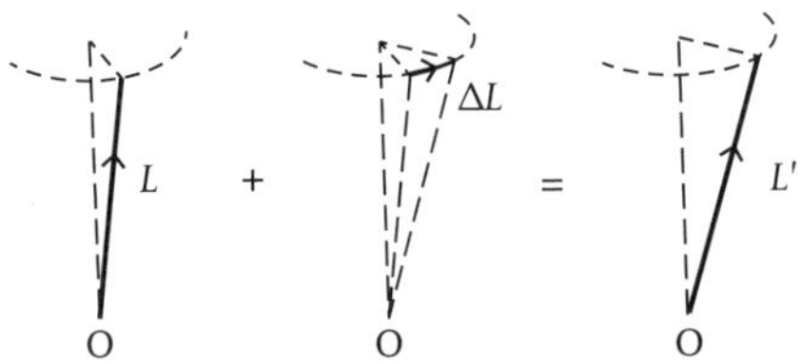

Fig. 165 Explanation for the precession of a spinning top.

For the torque vector due to gravity is perpendicular to the plane containing the top's axis and the vertical through O, leading, a short time Δt later, to an extra angular momentum $\Delta\boldsymbol{L}$ in the direction shown. By the laws of vector addition, this produces the new angular momentum vector $\boldsymbol{L}'$, thereby accounting for the new position of the top's axis.

* * * * * * *

There is an interesting chapter on Sophie Kowalevski's life and work in

The Secret Lives of Numbers, by Kate Kitagawa and Timothy Revell (Viking, 2023).

More details about the Kowalevski top of Fig. 63b, including video clips of its motion, can be found in a talk by Andreas Malmendier at https://www.youtube.com/watch?v=v1Ro1gYCrdY.

* * * * * * *

A demonstration of the *chiral* tippy-top in Fig. 64 can be found towards the end of Tadashi Tokieda's 2008 LMS Popular Lecture at https://www.youtube.com/watch?v=pkfDYOZ1p4Y.

18. ...and Falling Cats

Much has been written on the falling cat problem, but see, in particular

Why Cats Land on their Feet, by Mark Levi (Princeton University Press, 2012);

Falling Felines and Fundamental Physics, by Gregory J. Gbur (Yale University Press, 2019).

19. The Not-so-Simple Pendulum

The 'proposition well known to Geometers' in connection with Fig. 70 can be proved as follows.

First, the purely geometrical result in Fig. 166 can be established by joining O to the mid-point of the chord and using similar right-angled triangles to show that $h/D = D/2r$.

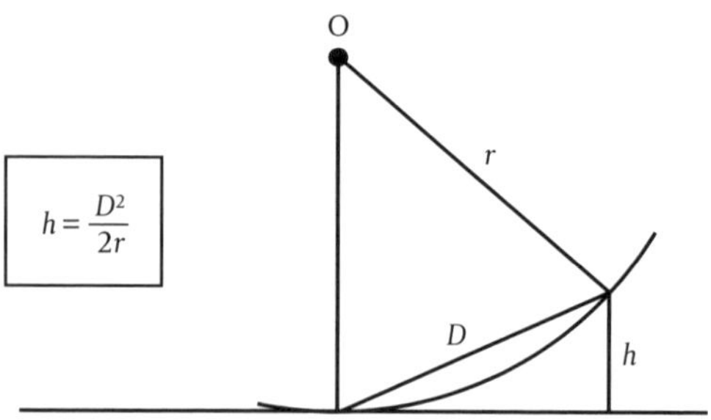

Fig. 166 Geometry of a large pendulum swing.

Now let the pendulum be released from rest, and have speed v at the lowest point. Equating the loss of potential energy with the gain in kinetic energy then gives

$$\frac{1}{2}mv^2 = mgh.$$

As h is proportional to D^2, and r is a constant, it follows that v is proportional to D.

* * * * * * *

For a full discussion (and translation) of Daniel Bernoulli's multiple-pendulum paper of 1738 see

The Evolution of Dynamics: Vibration Theory from 1687 to 1742, by J. T. Cannon and S. Dostrovsky (Springer-Verlag, 1981).

For a general history of pendulum dynamics see

The Pendulum Paradigm, by Martin Beech (Brown Walker Press, 2014).

20. The Wave Equation

As a simple example of *partial differentiation*, suppose that some quantity z depends on two independent variables x and t in such a way that $z = x^2t + t^2$. Then, in this particular case,

$$\frac{\partial z}{\partial x} = 2xt, \quad \frac{\partial z}{\partial t} = x^2 + 2t.$$

* * * * * * *

To establish the frequency formula relating to Fig. 76 from the wave equation in Fig. 73 we may proceed as follows.

Seek a solution $y = f(x) \sin \omega t$, so that all parts of the string oscillate with the same angular frequency ω. This leads to the ordinary differential equation

$$T\frac{d^2f}{dx^2} + \rho\omega^2 f = 0.$$

Solutions of this which have $f(0) = f(l) = 0$, so that the displacement at the two ends of the string is always zero, are

$$f(x) = D\sin\frac{N\pi x}{l},$$

for integer values of N, provided that

$$\omega = \frac{N\pi}{l}\sqrt{\frac{T}{\rho}}.$$

The actual frequency (number of oscillations per unit time), $\omega/2\pi$, is then given by the formula accompanying Fig. 76.

21. 'The Lawes . . . of Springinesse'

For a good biography of Robert Hooke see

The Man who Knew Too Much, by Stephen Inwood (Macmillan, 2002),

from which Fig. 78 is in fact taken.

For more on the history of elasticity see

'Outline of the History of Flexible or Elastic Bodies to 1788', by C. Truesdell, in *J. Acoust. Soc. Am.* **32** (1960), 1647–56;

The Rational Mechanics of Flexible or Elastic Bodies 1638–1788, by C. Truesdell (Orell Füssli, 1960);

History of Strength of Materials, by S. P. Timoshenko (Dover, 1988).

22. The Best of All Possible Worlds?

For very readable accounts of maximum and minimum principles more generally see

The Parsimonious Universe, by S. Hildebrandt and A. Tromba (Copernicus, 1996);
Reaching for the Extreme, by Ian Stewart (Profile Books, 2026).

Rather more technical accounts can be found in

Perfect Form, by Don S. Lemons (Princeton University Press, 1997),
The Best of All Possible Worlds, by Ivar Ekeland (University of Chicago Press, 2006).

* * * * * * *

A simple example of the Lagrange equations in Fig. 84 is provided by an object falling freely under gravity. Only one coordinate q is then needed to specify its position, and the obvious choice is the height y. Then $L = \frac{1}{2}m\dot{y}^2 - mgy$, and, in consequence,

$$\frac{\partial L}{\partial \dot{y}} = m\dot{y}, \quad \frac{\partial L}{\partial y} = -mg,$$

whereupon the (single, in this case) Lagrange equation reduces to

$$m\frac{d^2y}{dt^2} = -mg,$$

just as it should.

23. Instability

The spinning box of Fig. 85, and several other aspects of instability, are discussed further in my book *From Calculus to Chaos* (Oxford University Press, 1997).

More on the dynamics of the rattleback can be found in

'Celt Reversals: A Prototype of Chiral Dynamics', by H. K. Moffatt and T. Tokieda, *Proc. Roy. Soc. Edinburgh*, **138A** (2008): 361–8.

24. Fluid Dynamics

Three of the best general sources for the history of fluid dynamics are

Worlds of Flow: A History of Hydrodynamics from the Bernoullis to Prandtl, by Olivier Darrigol (Oxford University Press, 2005);
Rational Fluid Mechanics, 1687–1765, by Clifford Truesdell (Orell Füssli, 1954);
History of Hydraulics, by H. Rouse and S. Ince (Dover, 1963).

All three references include good discussions of what Daniel Bernoulli actually did in connection with what is now called 'Bernoulli's theorem'.

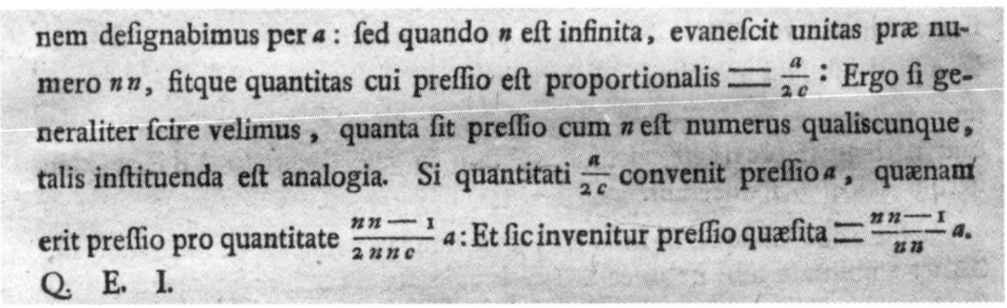
nem deſignabimus per a : ſed quando n eſt infinita, evaneſcit unitas præ numero nn, fitque quantitas cui preſſio eſt proportionalis $= \frac{a}{2c}$: Ergo ſi generaliter ſcire velimus, quanta ſit preſſio cum n eſt numerus qualiscunque, talis inſtituenda eſt analogia. Si quantitati $\frac{a}{2c}$ convenit preſſio a, quænam erit preſſio pro quantitate $\frac{nn-1}{2nnc}$ a: Et ſic invenitur preſſio quæſita $= \frac{nn-1}{nn}$ a. Q. E. I.

Fig. 167 The short passage from Daniel Bernoulli's *Hydrodynamica* (1738) most closely related to his famous theorem.

(Note that in my own statements of that theorem in Chapter 24, p denotes the pressure *due to the fluid motion*, i.e. the actual pressure less the hydrostatic pressure due to gravity.)

* * * * * * *

Figure 168 shows Euler's original statement of the general equations of nonviscous flow, as they appear in the *Mémoires de l'Académie des Sciences* (Berlin, 1755).

XV. **Maintenant je poſe pour abréger:**

$$X = \left(\frac{du}{dt}\right) + u\left(\frac{du}{dx}\right) + v\left(\frac{du}{dy}\right) + w\left(\frac{du}{dz}\right)$$

$$Y = \left(\frac{dv}{dt}\right) + u\left(\frac{dv}{dx}\right) + v\left(\frac{dv}{dy}\right) + w\left(\frac{dv}{dz}\right)$$

$$Z = \left(\frac{dw}{dt}\right) + u\left(\frac{dw}{dx}\right) + v\left(\frac{dw}{dy}\right) + w\left(\frac{dw}{dz}\right),$$

& l'équation différentielle qui détermine la preſſion p eſt:

$$\frac{dp}{q} = P\,dx + Q\,dy + R\,dz - X\,dx - Y\,dy - Z\,dz,$$

dans laquelle le tems t eſt ſuppoſé conſtant. Or l'autre équation tirée de la continuité du fluide eſt:

$$\left(\frac{dq}{dt}\right) + \left(\frac{d.qu}{dx}\right) + \left(\frac{d.qv}{dy}\right) + \left(\frac{d.qw}{dz}\right) = 0,$$

& ce ſont les deux équations qui contiennent toute la Théorie tant de l'équilibre que du mouvement des fluides, dans la plus grande univerſalité qu'on puiſſe imaginer.

Fig. 168 The first appearance of the general equations of nonviscous fluid mechanics (1755).

The 'curly' symbols for partial differentiation are not yet in use, (X, Y, Z) denote the components of fluid acceleration, and (P, Q, R) denote the components of body force per unit mass, which most usually will be $(0, 0, g)$ due to gravity.

Euler uses q for the fluid density while we now use ρ, and the final equation, expressing conservation of mass, is typically written today in vector calculus form as

$$\frac{\partial \rho}{\partial t} + \nabla \cdot (\rho \boldsymbol{u}) = 0,$$

where $\boldsymbol{u}$ denotes the fluid velocity. For an incompressible fluid of constant density ρ it reduces to

$$\nabla \cdot \boldsymbol{u} = 0.$$

25. Viscous Flow

In his treatment of viscous flow outside a rotating cylinder (Fig. 93b), Newton concludes that the rotary flow velocity v_θ is independent of the radial coordinate r, while it is actually proportional to $1/r$.

The error arises because he applies what we now call the principle of linear momentum when what his line of argument actually needed is the principle of angular momentum, only clarified in its full generality by Euler in 1775.

The error was first identified by Stokes in pp. 303–4 of his classic 1845 paper on the fundamental equations of viscous flow in *Trans. Camb. Phil. Soc.*, **8**: 287–319.

* * * * * * *

In the notation of modern vector calculus, the Navier–Stokes equations of motion for an incompressible fluid, of constant density ρ and viscosity μ, are

$$\rho\left(\frac{\partial \boldsymbol{u}}{\partial t} + \boldsymbol{u}\cdot\nabla\boldsymbol{u}\right) = -\nabla p + \rho\mathbf{g} + \mu\nabla^2\boldsymbol{u},$$

$$\nabla\cdot\boldsymbol{u} = 0.$$

When written out in full component form, they amount to four scalar partial differential equations for four 'unknowns', namely the three components of fluid velocity $\boldsymbol{u}$ and the pressure p.

The first equation above—called the momentum equation—results from *force* = *mass* × *acceleration* applied to a small element of fluid. The quantity in brackets on the left-hand side is the fluid element's acceleration, which is *not* simply $\partial\boldsymbol{u}/\partial t$, because that would represent the rate of change of $\boldsymbol{u}$ with time at *a fixed position in space*, rather than for a particular (and moving) fluid element.

The second equation expresses the incompressibility of the fluid, and reduces to the one shown in Fig. 91 when the flow is two-dimensional.

* * * * * * *

For more on the Reynolds number see

Note on the History of the Reynolds Number, by N. Rott, *Ann. Rev. Fluid Mech.*, **22** (1990): 1–11.

26. 'I Got Instantly on Horseback'

For more on the history of water waves see

Worlds of Flow, by Olivier Darrigol (Oxford University Press, 2005);
'The Origins of Water Wave Theory', by A. D. D. Craik, in *Ann. Rev. Fluid Mech.*, **36** (2004): 1–28.

* * * * * * *

If y denotes the upward displacement of the free surface, and h the undisturbed depth, the Korteweg–de Vries equation is

$$\frac{\partial y}{\partial t} + c\frac{\partial y}{\partial x} + \frac{3c}{2h}y\frac{\partial y}{\partial x} + \frac{ch^2}{6}\frac{\partial^3 y}{\partial x^3} = 0,$$

where $c = \sqrt{gh}$. It is valid when $y \ll h \ll L$, where L is a characteristic horizontal length scale of the wave.

The solitary wave solution is

$$y = a\,\mathrm{sech}^2\left\{\sqrt{\frac{3a}{4h^3}}\,(x - Vt)\right\},$$

where

$$V = c\left(1 + \frac{a}{2h}\right).$$

27. Shock Waves

The differential equation which Stokes was studying in connection with Fig. 105 is

$$\frac{\partial u}{\partial t} + (c+u)\,\frac{\partial u}{\partial x} = 0,$$

where c denotes the speed of small-amplitude sound waves. The general (but implicit) solution is

$$u = f[x - (c+u)\,t]\,,$$

where f denotes an arbitrary function of one variable.

It follows that any particular value of the flow velocity u travels in the positive x-direction at speed $c + u$. In a typical small-amplitude sound wave, u is negligible compared to c, all values of u travel at the same speed c, and the waveform travels without change of shape. But when u is too large to be neglected in comparison with c, the fact that larger values of u then travel faster results in the wave steepening of Fig. 105.

28. Vortex Motion

Much more on vortex motion can be found in my book *Elementary Fluid Dynamics* (Oxford University Press, 1990).

For a very wide-ranging and less mathematical look at vortex motion see

Vortex Flow in Nature and Technology, by Hans Lugt (Wiley, 1983).

29. The Curse of the Golden Ratio

A video of the 'walkabout' instability of vortex leapfrogging in Fig. 114 can be seen on YouTube at https://www.youtube.com/watch?v=PUkGD_RRCYo

My original paper on the subject was published as 'Instability of Vortex Leapfrogging' in *Eur. J. Phys.*, **21** (2000): 269–73.

For a very readable but balanced account of the golden ratio, see

The Golden Ratio, by Mario Livio (Headline Review, 2002).

30. The Coriolis Force

For a thought-provoking essay on the history of the Coriolis force see

'The Coriolis Effect: Four Centuries of Conflict between Common Sense and Mathematics, Part 1: A History to 1885', by A. O. Persson, in *History of Meteorology*, **2** (2005): 1–24.

31. The Three-Body Problem

A simple example of Hamilton's equations in action is again provided by a constant mass m falling under gravity, and may be compared with the Lagrange formulation in the 'Notes' for Chapter 22, p. 243.

In this case $q = y$, the upward momentum $p = mdy/dt$, and the Hamiltonian H is the total energy:

$$H = \frac{p^2}{2m} + mgq.$$

The Hamilton equations in Chapter 31 then reduce to

$$\frac{dq}{dt} = \frac{p}{m}, \quad \frac{dp}{dt} = -mg,$$

which together imply, yet again:

$$m\frac{d^2y}{dt^2} = -mg.$$

* * * * * * *

For more on the three-body problem and chaos see

Poincaré and the Three Body Problem, by June Barrow-Green (American Mathematical Society, 1991);
Celestial Encounters, by Florin Diacu and Philip Holmes (Princeton University Press, 1996).

32. A Paradox Resolved

A remarkable history of aerodynamics, including first-hand accounts of working with Ludwig Prandtl, can be found in

Aerodynamics: Selected Topics in the Light of Their Historical Development, by Theodore von Kármán (Dover, 2004).

33. The Secrets of Flight

For very readable and lightly mathematical accounts see

Mastery of the Air: An Account of the Science of Mechanical Flight, by Sir Graham Sutton (Hodder and Stoughton, 1965);
The Miracle of Flight, by Stephen Dalton (McGraw-Hill, 1977).

The second of these, from which Fig. 134 was taken, has an emphasis on the flight of birds and insects.

* * * * * * *

One crucial aspect of the circulation theory of lift is that the main flow is 'irrotational'. In the language of vector calculus, this means that $\nabla \wedge \boldsymbol{v} = 0$, where $\boldsymbol{v}$ denotes the fluid velocity. And, as it happens, this provides an alternative explanation for why $p + \frac{1}{2}\rho v^2$ is constant everywhere, rather than just along an individual streamline.

The precise definition of circulation round a closed circuit C drawn in the fluid is the line integral

$$\oint_C \mathbf{v}.d\mathbf{x}.$$

And the reason that the starting vortex causes the establishment of an equal but opposite circulation round the wing itself is a major result in nonviscous fluid mechanics called Kelvin's circulation theorem.

A full exposition of all these subtleties can be found in Chapter 1 of my book *Elementary Fluid Dynamics* (Oxford University Press, 1990).

34. Falling through the Earth

An extensive history of this whole topic can be found in

Going Underground: The Science and History of Falling through the Earth, by Martin Beech (World Scientific, 2019).

35. Tremors in the Night

For a good general introduction to elasticity and elastic waves see

An Introduction to the Properties of Condensed Matter, by D. J. Barber and R. Loudon (Cambridge University Press, 1989),

from which the quote on p. 198–9 is in fact taken.

For more about seismic waves and the Earth's deep interior I recommend starting here: https://en.wikipedia.org/wiki/Internal_structure_of_Earth

36. Very Viscous Flow

The experiment in Fig. 145 can be seen in G. I. Taylor's classic film *Low Reynolds Number Flows*, which can be found on YouTube, with the particular sequence of interest featuring between 13:13 and 14:54.

A famous article on the swimming of microscopic organisms is

'Life at low Reynolds Number', by E. M. Purcell, in *Am. J. Phys.*, **45** (1977): 1–11,

though recent papers that I have seen indicate that the whole subject is still in a state of flux.

Hele-Shaw's original article appeared in *Nature*, **58** (1898): 34–6.

37. Which Way Does the Bathwater Spin?

The original article on this question is

'Bath-Tub Vortex', by A. H. Shapiro, in *Nature*, **196** (1962): 1080–1,

and the sequel appeared three years later:

'The Bath-Tub Vortex in the Southern Hemisphere', by L. M. Trefethen et al., in *Nature*, **207** (1965): 1084–5.

A video clip of the original experiment can be seen from 20:14 to 22:30 in a classic educational film from the 1960s called *Vorticity*, which can be found, for example, at https://www.youtube.com/watch?v=loCLkcYEWD4

* * * * * * *

Relative to a reference frame rotating with angular velocity Ω, the Navier–Stokes equations need to be modified in two ways.

First, p has a slightly different meaning, owing to centrifugal effects. Much more importantly, however, a Coriolis term $2\Omega\wedge\mathbf{u}$ must be added to the acceleration on the left-hand side.

The classic text on rotating fluids, from which Fig. 151b is taken, is

The Theory of Rotating Fluids, by H. P. Greenspan (Cambridge University Press, 1968).

For a more recent and wide-ranging account of the theory of rotating flows see

The Dynamics of Rotating Fluids, by P. A. Davidson (Oxford University Press, 2024).

* * * * * * *

The Taylor–Proudman theorem is as follows.

Let (u, v, w) be the flow velocity components relative to a frame rotating about the z-axis with angular velocity Ω.

When the conditions for the Taylor–Proudman theorem apply, the Coriolis force, which has no z-component, is the only one that can balance the pressure gradient:

$$
\begin{aligned}
-2\rho\Omega v &= -\frac{\partial p}{\partial x}, \\
2\rho\Omega u &= -\frac{\partial p}{\partial y}, \\
0 &= -\frac{\partial p}{\partial z}.
\end{aligned}
$$

The pressure p is therefore independent of z, so $\frac{\partial p}{\partial x}$ and $\frac{\partial p}{\partial y}$ are also independent of z. In consequence, u and v are independent of z.

Moreover, combining the first two equations with the incompressibility condition

$$\frac{\partial u}{\partial x} + \frac{\partial v}{\partial y} + \frac{\partial w}{\partial z} = 0$$

leads to $\frac{\partial w}{\partial z} = 0$, so that w also is independent of z.

The whole flow, relative to the rotating frame, is therefore *the same in all planes perpendicular to the rotation axis.*

* * * * * * *

As so often in mechanics, the precise history of this extraordinary result is both messy and complex. Geoffrey Taylor's (largely experimental) paper was published in 1923, while Joseph Proudman's theoretical study was somewhat earlier, in 1916, and very largely about rather different matters.

Yet the result itself, essentially derived as above, can be found on p. 208 of a paper in 1897 by S. S. Hough in *Proc. Roy. Soc. Lond. A*, **189**, 201–57. Moreover, the particular aspect $\frac{\partial w}{\partial z} = 0$ was glimpsed by Lord Kelvin at least as early as 1868 in a little-known experiment which is described, along with much other relevant history, in

'Kelvin's Discovery of Taylor Columns', by O. U. Velasco Fuentes, in *Eur. J. Mech. B/Fluids*, **28** (2009): 469–72.

38. Tipping Points, Catastrophe, and Chaos

For more on tipping points and catastrophe theory see

An Introduction to Catastrophe Theory, by P. T. Saunders (Cambridge University Press, 1980);
Climate, Chaos and Covid, by Chris Budd (World Scientific, 2023).

Two classic 'popular' accounts of chaos are

Chaos, by James Gleick (Abacus, 1993);
Does God Play Dice?, by Ian Stewart (Penguin, 1990);

but see also

The New Scientist Guide to Chaos, edited by N. Hall (Penguin, 1991),
The Essence of Chaos, by E. N. Lorenz (University College London Press, 1993).

Lorenz's classic paper 'Deterministic Nonperiodic Flow' is in *J. Atmos. Sci.*, **20** (1963): 130–41.

A more extensive account of D. W. Allan's disk-dynamo work can be found in *Math. Proc. Camb. Phil. Soc.*, **58** (1962): 671–93.

The paper by K. A. Robbins is in the same journal, **82** (1977): 309–25.

39. And Finally . . .

Figure 159 was created from an applet on a website called *Moonshot* at

https://math.sun.ac.za/bbartlett/assets/moonshot.html

which also contains much other information about the Apollo missions.

* * * * * * *

The instability in Fig. 161 is known as the Saffman–Taylor instability, and their original paper was published in *Proc. Roy. Soc. Lond. A*, **245** (1958): 312–29.

* * * * * * *

Full details of the upside-down pendulum theorem in Fig. 161 can be found in my article 'A Pendulum Theorem', in *Proc. Roy. Soc. Lond. A*, **443** (1993): 239–45.

Tom Mullin and I published the experimental results in *Nature*, **366** (1993), 215–16. We noted there one puzzling lack of agreement between the theorem and the experiment, and this eventually turned out to be due to a somewhat naive 'modelling' error on my part in which I had overlooked the (substantial) mass of the bearings which joined the rods. The revised comparison between the theorem and the inverted triple pendulum experiment of Fig. 162b is shown in Fig. 169.

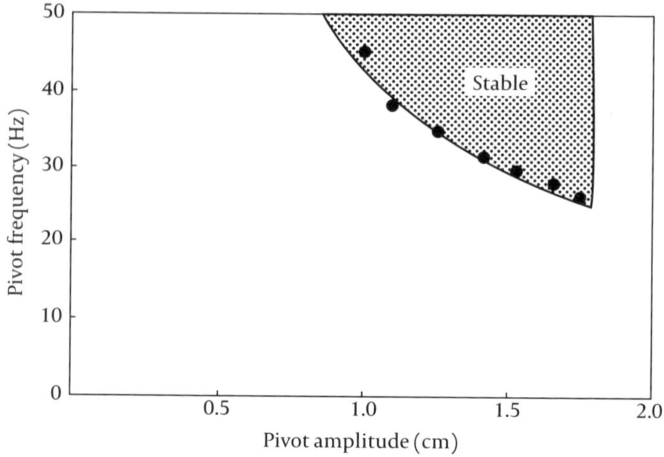

Fig. 169 Comparison between theory and experiment for the upside-down triple pendulum. Each of the three pendulums has length 19 cm, and each of the two bearings linking them has a mass almost twice that of a pendulum.

Further details can be found in Chapter 12 of my book *From Calculus to Chaos* (Oxford University Press, 1997).

Acknowledgements

I wish to thank Dan Taber, Giulia Lipparini, and, most especially, Emma Slaughter of Oxford University Press for all their invaluable help with this book.

I should also like to take this opportunity to thank John Johnson and Norman Riley for their inspirational help when I was first grappling with fluid mechanics as a research student at the University of East Anglia, some 50 years ago.

Figure Acknowledgements

Figure 1	Getty Images/Popperfoto
Figure 2a	Godfrey Kneller, public domain, via Wikimedia Commons
Figure 2b	Getty Images/Bettmann
Figure 4	Getty Images/Bettmann
Figure 6	Getty Images/ZU_09
Figure 10	Adapted from Getty Images/Bettmann
Figure 12a	Public domain, via Wikimedia Commons
Figure 20	Getty Images/Heritage Images
Figure 26	Giovanni Battista Landini, public domain, via Wikimedia Commons
Figure 43	Adapted from Getty Images/Bettmann
Figure 47a	Vasilij Sokolov, public domain, via Wikimedia Commons
Figure 51	Claude Francis Milliet Dechales, public domain, via Wikimedia Commons
Figure 60	Erik Gustafson, public domain, via Wikimedia Commons
Figure 63a	Public domain, via Wikimedia Commons
Figure 66	Étienne-Jules Marey, public domain, via Wikimedia Commons
Figure 67	Juniors Bildarchiv/juniors@wildlife Bildagentur GmbH/SuperStock
Figure 71a	By Unknown, edited by Bammesk - Basel Historical MuseumLink: https://www.hmb.ch/en/museums/objects-in-the-collection/details/s/portraet-des-daniel-bernoulli/, CC BY-SA 4.0, https://commons.wikimedia.org/w/index.php?curid=114825929
Figure 80	Public domain, via Wikimedia Commons
Figure 83a	Public domain, via Wikimedia Commons
Figure 83b	Nicolas de Largillière, public domain, via Wikimedia Commons
Figure 84b	Public domain, via Wikimedia Commons
Figure 95	Public domain, via Wikimedia Commons
Figure 101	Getty Images/Jeff J. Mitchell

Figure 115c Alamy Stock\Ian Dagnall Computing
Figure 116 Alamy Stock\The Print Collector
Figure 123 Public domain, via Wikimedia Commons
Figure 143 The Structure of the Earth on page 7 of the article produced by the Seismograph station operated by the British Geological Survey. Figure reproduced with permission from the British Geological Survey © UKRI 2025, all rights reserved. Available: https://earthquakes.bgs.ac.uk/education/eq_guide/earthquake_booklet.pdf. Permit Number CP25/037
Figure 144 Seismogram recorded in the UK from a distant earthquake on page 7 of the article produced by the Seismograph station operated by the British Geological Survey. Figure reproduced with permission from the British Geological Survey © UKRI 2025, all rights reserved. Available: https://earthquakes.bgs.ac.uk/education/eq_guide/earthquake_booklet.pdf. Permit Number CP25/0 37
Figure 151b Reproduced with permission via PLS Clear from H. P. Greenspan, *Theory of Rotating Fluids* (Cambridge University Press 1968)
Figure 157b Robbins K. A., 'A new approach to subcritical instability and turbulent transitions in a simple dynamo'. *Mathematical Proceedings of the Cambridge Philosophical Society*. 1977; 82(2): 309–325, Reproduced with permission

Index

A

acceleration:
  constant 10
  defined 19
  due to gravity 11
  in circular motion 20
algebra 34
angular momentum 82
angular velocity 71, 83
  of Earth 212
Allan, D. W. 220
area law, Kepler's 42, 54, 72

B

bathtub vortex 209
bending stiffness 113, 124
Bernoulli, Daniel 102
  double pendulum 102, 221, 226
  theorem in fluid mechanics 127, 184, 244
Bernoulli, James 112
Bernoulli, Johann 67
Bohr, Niels 87, 109, 158
boundary layer 179
  separation 180
buckling, elastic 124
bulk modulus 200

C

calculus 57
  of variations 119
cat, falling 93
catastrophe 216
Cauchy, Augustin-Louis 199
centre of gravity 84
central force 53
centrifugal force 167
centrifugal railway 23
centripetal force 21, 70
chaos 218
  in three-body problem 175
Chladni, Ernst 115
circular motion 18, 44
circulation:
  round a wing 186, 251
  secondary 182
clap-and-fling mechanism 188
Collignon, Édouard 193
collisions 28
components:
  of force 13
  of velocity 69, 82
conical pendulum 43
conservation:
  of angular momentum 84
  of energy 25, 29, 240
  of momentum 29
conservation law 121
coordinates:
  generalised 173
  polar 69
  rectangular 68
Coriolis force 168, 211

D

D'Alembert, Jean le Rond:
- flow past a sphere 131
- paradox 132, 140, 178, 181
- wave equation 105

density 127
Descartes, René 133
differential equations 62, 64, 129
- for point mass 68
- for simple pendulum 100

differentiation 58
- partial 105

dimensions, method of 36
dispersion 144
double pendulum
- and chaos 222
- of Bernoulli 102

dynamo 220

E

Earth:
- interior 195, 202
- liquid core 202
- mass 195
- radius 195
- rotation of 73, 209

earthquake 198
Eiffel tower 76
elastica 112
elasticity:
- buckling 124
- catastrophe 217
- Hooke's law 111, 199
- linear 199

electromagnetic wave 106
ellipse 40, 55
Encarsia formosa 188
energy 23
- conservation 25
- in Lagrange's mechanics 119
- kinetic 24
- potential 25

Euclid 50
Euler, Leonhard 3, 67
- and Newton 57
- angular momentum 84, 236
- applying calculus to dynamics 68
- buckling 124
- calculus of variations 119
- elasticity 113
- fluid motion 129, 136, 245
- linear momentum 77, 235
- planetary motion 71
- wave reflection 107

F

falling:
- object 12, 17, 59, 73
- through the Earth 190

Fermat, Pierre de 116
Fibonacci sequence 164
fish propulsion 206
Flammarion, Camille 190
flow:
- compressible 148
- high Reynolds number 139, 177
- incompressible 129
- low Reynolds number 205
- past circular cylinder 178
- past sphere 131
- past wing 183
- pipe 137
- turbulence 139
- vortex 154, 160

fluid mechanics:
- non-viscous 130, 141, 148, 154, 160, 209
- of Daniel Bernoulli 127
- of Euler 129, 245
- viscous 133, 159, 179, 187, 204, 225

focal points 41
force:
- as a vector 13

force (*Continued*)
  centrifugal 167
  centripetal 21
  due to gravity 12, 194
Foucault pendulum 166

G

Galileo 5
  chord theorem 16, 231
  inclined plane 6
  Leaning Tower of Pisa 5, 229
  on relativity 31
  parabolic motion 17
  falling through the Earth 192
geometry:
  in Newton's *Principia* 50
Golden Ratio 164
gravitation:
  inside the Earth 196
  inverse square law 194
  slingshot 223
  three-body problem 171, 224
gravity, centre of 84, 88
guitar string vibrations 108
Gulf Stream 215
Guyou, Émile 95

H

Halley, Edmund:
  and algebra 35
  and Newton 45
Hamilton's equations 173, 249
Hele-Shaw cell 207
Helmholtz, Hermann von 155
Hermann, Jacob 61
Hobbes, Thomas 35
Hooke, Robert:
  elasticity, law of 63, 111
  letter to Newton 45
  and pendulums 42
Huygens, Christiaan:
  centripetal acceleration 43
  collisions 29
  pendulum 99
hysteresis 217

I

images, method of 107
impulse 48, 53
inclined plane experiments 6, 15, 229
incompressibility equation 129
initial conditions 64, 222
insect flight 189
instability:
  buckling 124
  Reynolds' experiment 139
  spinning box 123
  three-body problem 172
  very viscous flow 225
  vortex leapfrogging 162
inverse square law 56, 194
isobars 214

J

Jeffreys, Sir Harold 198
Joukowski, Nikolai 186
jumps 216

K

Kelvin, Lord 155
Kepler, Johannes 41
  equal area law 42
kinetic energy 24
Korteweg de Vries equation 146, 247
Kowalevski, Sophie 90
Kutta, Martin 185

L

Ladies Diary, The 192
Lagrange's equations 120, 243
laminar flow 139
Lanchester, Frederick 185

Laplace, Pierre-Simon 148
Leaning Tower of Pisa 5, 12, 59, 229
leapfrogging 156, 161
least action, principle of 117
least time, principle of 116
Leibniz, Gottfried 58, 117
lever, law of 81
lift on a wing 183
light:
  electromagnetic nature 106
  refraction of 116
linearity 107
loop-the-loop 24
Lorenz, Ed 218
Love, Augustus 161

M

Mach number 149
magnetic field 220
Marci, Marcus 28
mass 2, 12, 24, 28
Maupertuis, Pierre de 117
'Maths Inspiration' 122
Maxwell, James Clerk 106
meteorology 214, 219
method of images 107
mode of vibration 103, 108
moment:
  of a force 81
  of inertia 83
  of momentum 82
momentum:
  angular 82, 84
  generalised 173
  linear 28, 77
Moon missions 223
Moore, C. 224
Mullin, T. 226
multiple states 217

N

Navier-Stokes equations 135, 246, 252
Newton, Sir Isaac 2
  collisions 49
  gravitation 194
  laws of motion 48
  planetary motion 52
  Principia 47
  three-body problem 171
  viscous flow 133
  water waves 144
nodes 110
Noether, Emmy 120
nonlinearity 219
no-slip condition 136

O

oceanography 215
Oldham, R. 202
optics:
  algebra of 35
  least time, principle of 116
Oresme, Nicole 10
oscillation 63
  modes of 102, 108
  period 65
  of double pendulum 103
  of simple pendulum 99

P

partial differentiation 105, 129, 241
Pauli, Wolfgang 87
pendulum:
  chaos 222
  collisions 101
  conical 43
  double 102
  Foucault 166
  inverted 226, 255
  period 99
  simple 37, 100

phase space 173
Phoronomia 61
plane, inclined 6, 15
planetary motion:
  circular approximation 44
  elliptical orbit 41
  equal area law 42, 53
  Euler's treatment 70
  Kepler's laws 41
  Newton's treatment 52
  orbital times 42
Poincaré, Henri 175
Poisson, Siméon-Denis:
  elastic waves 201
  ratio 200
  water waves 143
polar coordinates 69
potential energy 25
Prandtl, Ludwig 177
precession 89, 239
predictability 219
pressure:
  in fluid mechanics 127
  gradient 136
Principia 47, 57
principle:
  of angular momentum 80, 84, 236
  of least action 117
  of least time in optics 116
  of linear momentum 77, 235
P waves 202

Q

quantum mechanics 109

R

Rademaker, G. G. J. 97
radial direction 69
radians 69
rate of change 58
rattleback 126
Rayleigh, Lord 153
reflection of waves 107
refraction of light 116
reversals of magnetic field 220
reversibility of very viscous flow 204
Reynolds' experiment 139
Reynolds number 137, 178
  and reversibility 205
  reduced 206
  typical values 138
rigid body motion 85, 237
Robbins, K. 221
rolling motion 15
rotating drum 1, 167
rotation:
  angular momentum 82
  angular velocity 71
  centripetal force 21
  in circle 18
  loop-the-loop 24
  of Earth 73, 209
  of reference frame 169, 211
  rigid body 85
  spinning top 89, 125
Royal Society 28, 42
Russell, John Scott 141

S

Schrödinger, Erwin 109
sensitivity to initial conditions 219, 222
separation 180
shear modulus 199
shear stress 200
shock wave 148, 151
simple harmonic motion 65
skater on ice 80
slingshot, gravitational 223
smoke ring 156
solitary wave 141
spacecraft trajectory 223

speed:
  constant 9
  different from velocity 19
  sound 148
spermatozoon 205
spider, toy 63, 174
spin-down of cup of tea 181
stability:
  of spinning top 125
  of inverted pendulums 227, 255
starting vortex 187
Stephenson, A. 226
Stevin, Simon 6, 14
Stokes, Sir George 135, 151, 201
streamlines 127, 131
string:
  vibrations of 109
  waves on 104
Sun, motion round 41
supersonic flow 150
S waves 201
symmetry 120

T

Taylor-Proudman theorem 212, 253
Ter Braak, J. W. G. 97
thermal convection 218
three-body problem 171, 223
  restricted 175, 224
  chaos in 176
tipping point 216
tippy-top 88, 91
  chiral 92
top, spinning 87
tornado 155
torque 80
tower problem 73, 235
trailing edge 184, 187
transverse direction 69
Truesdell, Clifford 236
tunnel through the Earth 190
turbulence 139

V

vector calculus 245
vector product 236
velocity:
  distinguished from speed 19
vibration:
  frequency 108
  modes of 109
viscosity, coefficient of 134
viscous flow:
  fingers 225
  in pipe 137
  reversibility 204
  stress in 134
Voltaire 118
vortex:
  bathtub 209
  interaction 159
  leapfrogging 156, 161
  motion 154
  pair 160
  ring 155
  starting 187
  theory of matter 157

W

Wallis, John 34
'walkabout' instability 163
water waves 143
wave equation 105
wavelength 143
waves:
  earthquake 201
  elastic 201
  on string 104
  reflection of 107
  seismic 201
  shock 151
  small-amplitude 143, 151
  solitary 145
  sound 149
  water 143
Weis-Fogh, Torkel 188